MANAGE INSECTS

ON YOUR FARM

A Guide to Ecological Strategies

Miguel A. Altieri and Clara I. Nicholls
with Marlene A. Fritz

Published by the Sustainable Agriculture Network
Beltsville, MD

Cover photos (clockwise from top):

Mixed annual clovers in a Chico, Calif., almond orchard help invigorate the crop's ability to withstand pests while attracting beneficial insects. Robert L. Bugg, UC-Davis.

Sevenspotted lady beetle. Russ Ottens, Univ. of GA.

ARS entomologist Marina Castelo Branco collects plant samples for a pollen library that documents plants that attract insect pests. Scott Bauer, USDA-ARS.

Syrphid fly. Vincent J Hickey, M.D.

Back cover: Predatory stink bug nymph on eggplant. Debbie Roos, North Carolina Cooperative Extension

Graphic design and layout: Andrea Gray

Project manager and editor: Andy Clark

This book was published by the Sustainable Agriculture Network (SAN) under cooperative agreements with the Cooperative State Research, Education, and Extension Service, USDA, the University of Maryland and the University of Vermont.

Every effort has been made to make this book as accurate as possible and to educate the reader. This text is only a guide, however, and should be used in conjunction with other information sources on pest management. No single pest management strategy will be appropriate and effective for all conditions. The editor/authors and publisher disclaim any liability, loss or risk, personal or otherwise, which is incurred as a consequence, directly or indirectly, of the use and application of any of the contents of this book.

SARE works to increase knowledge about — and help farmers and ranchers adopt — practices that are profitable, environmentally sound and good for communities. For more information about SARE grant opportunities and informational resources, go to www.sare.org. SAN is the national outreach arm of SARE. For more information, contact:

Sustainable Agriculture Network
10300 Baltimore Ave.
Bldg. 046 BARC-WEST
Beltsville, MD 20705-2350
(301) 504-5236; (301) 504-5207 (fax)
san_assoc@sare.org

To order copies of this book ($15.95 plus $5.95 s/h), contact (301) 374-9696, sanpubs@sare.org., or order online at www.sare.org.

Library of Congress Cataloging-in-Publication Data

Altieri, Miguel A.
  Manage insects on your farm : a guide to ecological strategies / by
Miguel A. Altieri and Clara I. Nicholls with Marlene Fritz.
      p. cm. -- (Sustainable Agriculture Network handbook series ; bk.
7)
  Includes index.
  ISBN 1-888626-10-0
  1. Agricultural pests--Biological control. 2. Insects as biological
pest control agents. 3. Agricultural ecology. 4. Ecological
engineering. I. Nicholls, Clara I. II. Fritz, Marlene, 1950- III.
Title. IV. Series.

SB975A48 2005
632'.96--dc22

2005013726

Acknowledgments

THIS BOOK COULD NOT HAVE BEEN PUBLISHED without the contributions of many scientists, educators and farmers. The concept for *Manage Insects on Your Farm: A Guide to Ecological Strategies* came out of a shorter bulletin, also published by SAN, titled, *A Whole Farm Approach to Ecological Pest Management.*

Miguel Altieri and Clara Nicholls, University of California-Berkeley, felt that the bulletin topic could be expanded, and authored this manuscript to explore the concept of ecological insect management in greater detail.

Marlene Fritz, University of Idaho Extension communications specialist, working with SAN staff, edited the manuscript, contacted numerous farmers, scientists and educators, wrote the farm features, and fleshed out the how-to sections of the book. Marlene also solicited and edited additional sections by experts in the other areas of ecological pest management.

Contributors:
Fred Magdoff, University of Vermont
Sharad Phatak, University of Georgia
John Teasdale, USDA-ARS, Beltsville, MD
Joe Lewis, University of Georgia
Glen Raines, University of Georgia
Luigi Ponti, University of California-Berkeley

The book was reviewed by the authors, the contributors and by numerous agriculturalists: Stefanie Aschmann, USDA-NRCS; Bob Bugg, University of California-Davis; Larry Dyer, Kellogg Biological Station; Lisa Krall,

USDA-NRCS; Doug Landis, Michigan State University; Tom Larson, St. Edward, NE; John Mayne, Southern SARE; Fabian Menalled, Iowa State University; Dale Mutch, Michigan State University; Debbie Roos, North Carolina Cooperative Extension; Kim Stoner, Connecticut Agricultural Experiment Station.

Numerous researchers, farmers and photographers worked with us to provide photos (see credits for individual photos). Special thanks to SARE program assistant Amanda Rodrigues for her research and organizational skills to pull these photos together.

SARE and SAN staff Valerie Berton, Andy Clark, Diana Friedman, Sarah Grabenstein, Kim Kroll and Amanda Rodrigues all contributed over the course of the project.

Andy Clark
Sustainable Agriculture Network
Beltsville, MD
August 2005

Contents

6 PUTTING IT ALL TOGETHER   86

RESOURCES   104

Index   115

1 Introduction

 AGRICULTURAL PESTS — insects, weeds, nematodes and disease pathogens — blemish, damage or destroy more than 30 percent of crops worldwide. This annual loss has remained constant since the 1940s, when most farmers and ranchers began using agrichemicals to control pests.

Agrichemical methods of protecting crops are costly to the farmer, potentially harmful to the environment and, despite widespread use, have not proved 100-percent effective. Problems persist due to pest resistance and the uncanny ability of pests to overcome single-tactic control strategies.

A National Academy of Science 1997 Proceedings paper, "A Total System Approach to Sustainable Pest Management," called for "a fundamental shift to a total system approach for crop protection [which] is urgently needed to resolve escalatory economic and environmental consequences of combating agricultural pests."

Many farmers are seeking such an approach, one that relies less on agrichemicals and more on mimicking nature's complex relationships among different species of plants and animals. Known as "ecologically based pest management" or simply "ecological pest management," this approach treats the whole farm as a complex system.

The old approach strives for 100 percent *control* of every pest using one strategy or agrichemical for each pest. The new approach, ecological pest management, aims to *manage* the whole farm and keep pests at acceptable populations using many complementary strategies. Ecological pest man-

A crimson clover cover crop prevents erosion, improves soil, fixes nitrogen and attracts beneficial insects.

agement is a *preventive* approach that uses "many little hammers" or strategies, rather than one big hammer, to address pest problems on the farm or ranch.

Ecological pest management employs tactics that have existed in natural ecosystems for thousands of years. Since the beginning of agriculture — indeed, long before then — plants co-evolved with pests and with the natural enemies of those pests. As plants developed inherent protective mechanisms against pests, they were helped by numerous partners in the ecosystem, for example:

- Beneficial insects that attack crop insects and mites by chewing them up or sucking out their juices
- Beneficial parasites, which commandeer pests for habitat or food
- Disease-causing organisms, including fungi, bacteria, viruses, protozoa and nematodes that fatally sicken insects or keep them from feeding or reproducing. These organisms also attack weeds.
- Insects such as ground beetles that eat weed seeds
- Beneficial fungi and bacteria that inhabit root surfaces, blocking attack by disease organisms

By integrating these natural strategies into your farming systems, you can manage pests in a way that is healthier for the environment and eliminates many of the problems associated with agrichemical use. Knowing the life cycles of pests and understanding their natural enemies allows you to better manipulate the system to enhance, rather than detract from, the built-in defenses available in nature. Another National Academy of Science report (1996), *Ecologically Based Pest Management* (EBPM), stated that EBPM "should be based on a broad knowledge

Aleiodes indiscretus wasp parasitizing a gypsy moth caterpillar.

of the agro-ecosystem and will seek to manage rather than eliminate pests" in ways that are "profitable, safe, and durable."

In addition to reducing pest damage, shifting your farming system to ecological pest management will bring multiple benefits to your operation. For example, moving from monoculture to longer rotations improves water- and nutrient-use efficiency. Cover crops planted to attract beneficial insects also suppress weeds, improve the soil, provide moisture-conserving mulch, fix or store nitrogen for subsequent crops and contribute to overall nutrient management goals.

About *Manage Insects on Your Farm*

Pests of agricultural crops include weeds, insects, pathogens and nematodes. This book is focused mostly on managing *insect* pests, but it addresses all crop pests to some degree, because no pest or category of pests can be addressed in isolation. The ecological pest management strategies presented here will contribute to overall ecosystem health.

We first lay out the principles behind ecologically based pest management. Then, we describe strategies used by farmers around the world to address insect problems within the context of their whole farm systems. A full section is devoted to how you can manage your soil to minimize insect damage. Flip to Chapter 5 to learn about beneficial insects you can put to work for you. Photos of some beneficials and pests can be found on pages 50–54.

COVER CROP SYSTEM DETERS PESTS

In Lancaster County, Pa., Steve Groff built a farming system based on cover crops, intensive crop rotation and no-till. Although he designed his crop and vegetable farm without targeting specific pests, Groff and the scientists using his farm as a real-world laboratory have documented significant benefits in pest management, including:

- Increased populations of beneficial insects in cover crops
- Reduced populations of Colorado potato beetles in tomatoes
- Delayed onset of early blight in tomatoes
- Minimal to no aphid pressure on any of his crops
- Reduced cucumber beetle damage in pumpkins
- Tolerable levels of European corn borer, thanks to releases of the parasitic wasp, *Trichogramma ostriniae*
- Reduced weed pressure, although monitoring and managing weeds are still a top priority on his farm

Those benefits come at some cost, however. Groff spends more time managing his complex system to ensure that cover crops are seeded and killed at the right time and to scout for weeds. Moreover, he monitors soil temperature because no-till and cover crop residues delay soil-warming in the spring.

Not all pest management problems have been solved, either. Spider mites still attack Groff's tomatoes, particularly in dry years, while slugs sometimes hide under cover crop residues in wet years. Nonetheless, consider the numbers. Groff has cut pesticide use by 40 percent and seen soil organic matter increase by almost 50 percent with a 10 percent net increase in yield averaged over all crops. "It's working for us," Groff says.

Groff's system is described in greater detail on pages 60–63.

R. Weil, Univ. of Maryland

Steve Groff's cover crop of cereal rye and flowering rapeseed provides multiple benefits compared to neighboring plowed fields.

Workers harvest celery next to a strip of bachelor button flowers planted to attract beneficial insects.

Throughout the book, we present specific examples of successful pest management strategies. While some examples may fit your farm or ranch, most are crop- or climate-dependent and will serve mostly to stimulate your imagination and help you better understand that while every system is unique, the general principles of ecological pest management apply universally. Use this book as a stepping-stone to develop a more complex, more diverse system on your own farm. Look for "Tip" boxes throughout the book for specific suggestions.

This book does not address the multiple ecological benefits of further diversifying your farm or ranch by integrating livestock into the system. If you also raise animals, consult other information resources about the management and benefits of integrated crop-livestock systems (Resources, p. 104).

In short, nature has already provided many of the tools needed to successfully combat agricultural pests. This book aims to describe those tools and present successful strategies for using them to manage insects on your farm or ranch.

2 How Ecologically Based Pest Management Works

 TO BRING ECOLOGICAL PEST MANAGEMENT to your farm, consider three key strategies:

- Select and grow a diversity of crops that are healthy, have natural defenses against pests, and/or are unattractive or unpalatable to the pests on your farm. Choose varieties with resistance or tolerance to those pests. Build your soil to produce healthy crops that can withstand pest pressure. Use crop rotation and avoid large areas of monoculture.
- Stress the pests. You can do this using various management strategies described in this book. Interrupt their life cycles, remove alternative food sources, confuse them.
- Enhance the populations of beneficial insects that attack pests. Introduce beneficial insects or attract them by providing food or shelter. Avoid harming beneficial insects by timing field operations carefully. Wherever possible, avoid the use of agrichemicals that will kill beneficials as well as pests.

EBPM relies on two main concepts:

Biodiversity in agriculture refers to all plant and animal life found in and around farms. Crops, weeds, livestock, pollinators, natural enemies, soil fauna and a wealth of other organisms, large and small, contribute to biodiversity. The more diverse the plants, animals and soil-borne organisms that inhabit a farming system, the more diverse the community of pest-fighting beneficial organisms the farm can support.

Biodiversity is critical to EBPM. Diversity, in the soil, in field boundaries, in the crops you grow and how you manage them, can reduce pest problems, decrease the risks of market and weather fluctuations, and eliminate labor bottlenecks.

Biodiversity is also critical to crop defenses: Biodiversity may make plants less "apparent" to pests. By contrast, crops growing in monocultures over large areas may be so obvious to pests that the plants' defenses fall short of protecting them.

Biological control is the use of natural enemies — usually called "beneficial insects" or "beneficials" — to reduce, prevent or delay outbreaks of insects, nematodes, weeds or plant diseases. Biological control agents can be introduced, or they can be attracted to the farming system through ecosystem design.

Naturally occurring beneficials, at sufficient levels, can take a big bite out of your pest populations. To exploit them effectively, you must:

1) identify which beneficial organisms are present;
2) understand their individual biological cycles and resource requirements; and
3) change your management to enhance populations of beneficials.

> *"It's a subtle effect, but over time the advantage increases.*
> *Your system moves slowly toward a natural balance*
> *and your pest problems decrease."*
> — ZACH BERKOWITZ, CALIFORNIA VINEYARD CONSULTANT

The goal of biological control is to hold a target pest below economically damaging levels — not to eliminate it completely — since decimating the population also removes a critical food resource for the natural enemies that depend on it.

In Michigan, ladybugs feed on aphids in most field crops or — if prey is scarce — on pollen from crops like corn. In the fall, they move to forest patches, where they hibernate by the hundreds under plant litter and snow. When spring arrives, they feed on pollen produced by such early-

BIOLOGICAL CONTROL VOCABULARY

When farmers release natural enemies, or **beneficials,** to manage introduced pests, they are using biological control tactics. *Classical* biological control is the importation and release of beneficial insects against exotic pests. When farmers add a species of natural enemy to a field where it is not currently present, or present only in small numbers, they are using *augmentation* biological control: they can either *inundate* a field with large numbers of natural enemies or *inoculate* it with relatively few at a critical time. When they conserve the augmented natural enemies or the ones that are already present in and around their fields, they are using *conservation* biological control. *Parasitoids* — a class of beneficials — are parasitic insects that kill their hosts.

Debbie Roos, North Carolina Cooperative Extension

Jack Kelly Clark, Univ. of Calif.

(above) Southern green stink bug eggs being parasitized by *Trissolcus basalis.*

(left) Assassin bug feeding on Colorado potato beetle larva.

season flowers as dandelions. As the weather warms, they disperse to alfalfa or wheat before moving on to corn. Each component of biodiversity — whether planned or unplanned — is significant. For example, if dandelions are destroyed during spring plowing, the ladybugs lose an important food source. As a result, the ladybugs may move on to greener pastures, or fail to reproduce, reducing the population available to manage aphids in your cash crop.

Research shows that farmers can indeed bring pests and natural enemies into balance on biodiverse farms by encouraging practices that build the greatest abundance and diversity of above- and below-ground organisms (Figure 1). By gaining a better understanding of the intricate relationships among soils, microbes, crops, pests and natural enemies, you can reap the benefits of biodiversity in your farm design. Further, a highly functioning diversity of

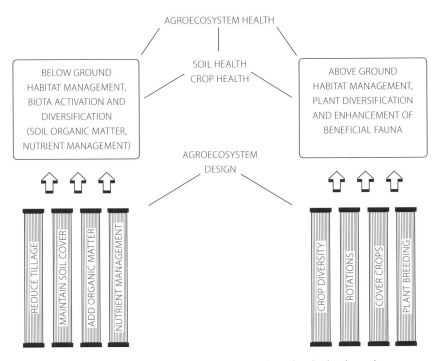

THE PILLARS OF ECOLOGICAL PEST MANAGEMENT

Figure 1. The pillars of ecological pest management, explained in this book, can be categorized into above-ground and below-ground principles and practices. Ecological pest management is based on the use of multiple tactics to *manage* pests in the agroecosystem, rather than a "silver bullet" to *control* them.

crucial organisms improves soil biology, recycles nutrients, moderates microclimates, detoxifies noxious chemicals and regulates hydrological processes.

What Does A Biodiverse Farm Look Like?

Agricultural practices that increase the abundance and diversity of above- and below-ground organisms strengthen your crops' abilities to withstand pests. In the process, you also improve soil fertility and crop productivity. Diversity on the farm includes the following components:

- Spatial diversity across the landscape (within fields, on the farm as a whole and throughout a local watershed)
- Genetic diversity (different varieties, mixtures, multilines, and local germplasm)
- Temporal diversity, throughout the season and from year to year (different crops at different stages of growth and managed in different ways)

How diverse is the vegetation within and around your farm?
How many crops comprise your rotation? How close is your farm
to a forest, hedgerow, meadow or other natural vegetation?
All of these factors contribute to your farm's biodiversity.

Ideally, agricultural landscapes will look like patchwork quilts: dissimilar types of crops growing at various stages and under diverse management practices. Within this confusing patchwork, pests will encounter a broader range of stresses and will have trouble locating their hosts in both space and time. Their resistance to control measures also will be hampered.

CAUTION! Increasing biodiversity takes a lot of knowledge and management, as it can backfire. Some cover crops can provide pest habitat, and mulches can boost populations of slugs, cutworms, squash bugs and other pests.

Plant diversity above ground stimulates diversity in the soil. Through a system of checks and balances, a medley of soil organisms helps maintain low populations of many pests. Good soil

Robert L. Bugg, Univ. of Calif.

A rosemary cash crop teams with flowering buckwheat, which improves the soil and attracts beneficials, in a Brentwood, Calif., apricot orchard.

YEAR-ROUND BLOOMING CYCLE ATTRACTS BENEFICIALS

In Oregon's Willamette Valley, Larry Thompson's 100-acre fruit and vegetable farm blossoms with natural insectaries. "To keep an equilibrium of beneficials and pests and to survive without using insecticides, we have as much blooming around the farm as we can," he says.

Thompson uses cover crops to recruit ladybugs, lacewings and praying mantises in his battle against aphids. Overseeded cereal rye is already growing under his lettuce leaves before he harvests in late summer and fall. "It creates a nice habitat for overwintering beneficials and you don't have to start over from ground zero in the spring," he says.

Between his raspberry rows, Thompson lets his dandelions flower into a food source for nectar- and pollen-seeking insects before mowing them down. Forced out of the dandelions that nurtured them in early spring, the beneficials pursue a succession of bloom. They move first into his raspberries, then his marionberries and boysenberries.

Later in the year, Thompson doesn't mow his broccoli stubble. Instead, he lets the side shoots bloom, creating a long-term nectar source into early winter. "The bees really go for that," he says.

Jerry DeWitt, Iowa State Univ.

The next generation of farmers? Students learn about ecological farm design from Oregon fruit and vegetable grower Larry Thompson.

tilth and generous quantities of organic matter also can stimulate this very useful diversity of pest-fighting soil organisms.

As a rule, ecosystems with more diversity tend to be more stable: they exhibit greater *resistance* — the ability to avoid or withstand disturbance — and greater *resilience* — the ability to recover from stress.

FARM FEATURE

DIVERSITY IN EVERY FIELD AND PEN

- Diversifies crops within space and time
- Plants windbreaks and grassy field borders
- Integrates crop and livestock operations
- Builds soils with diverse organic matter
- Uses resistant crops

It's been two decades since Ron and Maria Rosmann began transitioning their west central Iowa farm to organic. Their crops — soybeans, corn, alfalfa, turnips, grasses, oats, rye and other small grains — were certified organic in 1994. Their 90 stock cows and 650 broiler chickens followed in 1997, while their 20 antibiotic-free Berkshire sows are "natural pork."

Except for seed staining in their soybeans — transmitted by bean leaf beetles — and aphids and leafhoppers in their alfalfa, Rosmann Family Farms are bothered by few pests. While most of their neighbors have readily switched to "biotech" varieties, the Rosmanns' corn and soybean yields, over a 20-year average, are at least as high as the county's.

"Things are working well here and there's got to be a reason — and it's not just one," says Rosmann. "We look at it as a whole system."

Biodiversity is hard at work above and below ground

On their fourth-generation farm near Harlan, the Rosmanns plant windbreaks, grassy field borders and — for pheasants and quail — native prairie species. Generous populations of lacewings and ladybugs indicate that the Rosmanns' commitment to biodiversity is keeping predators in balance with prey. Nesting boxes support three pairs of American kestrels, which return the favor by snatching up small rodents.

Rather than alternating corn and soybeans every other year, the Rosmanns' primary rotation spans six years: corn, soybeans, corn, small grains and two years of alfalfa. Instead of expansive monocultures, they break up their 620 acres into about 45 fields, letting topography decide how each field is divided. If their light infestations of corn borers drop a few ears of corn onto the ground, their cattle glean them after harvest. "Most conventional farmers continue to tear out their fences," says Rosmann. "They don't have anything running on their fields to pick up the fallen grain. It's wasted on most farms. That's ridiculous."

Livestock enrich soils

If he had to offer just one reason why his farming system is so resilient, Rosmann would say it's his healthy soils. He beds his livestock in oat, rye and barley straw — his hogs are treated to the Swedish deep-bedding system of 2-foot-thick straw — then composts the straw with their manure. He feeds his soils every cubic inch of that compost and tills his fields very minimally. For example, he plants his corn and soybeans into ridges and turns those fields under only after the rotation's third year.

"I think our soil biology is balanced and that the bacteria, fungi and other microorganisms really help us out," he says. "They must be helping our productivity and breaking our disease and insect cycles."

Ron (left) and David Rosmann use long rotations and minimum tillage to grow healthy crops, resulting in minimal pest problems.

Indeed, the Rosmanns have only used one insecticide in their corn and soybeans in the past 20 years — *Bacillus thuringiensis* (Bt) against corn borers — but the insects didn't affect yields that year anyway and the Rosmanns haven't used the product since. "We try to keep our input costs down. As long as our yields are not being compromised, why purchase inputs?" he asks.

Rosmann controls the aphids and leafhoppers in his alfalfa by harvesting earlier when possible. That decreases production, but he can "put up with it." He also plants orchard grass with alfalfa, which discourages some pests.

Generous populations of lacewings and ladybugs
indicate that the Rosmanns' commitment to biodiversity
is keeping predators in balance with prey.

Besides soil health, the Rosmanns control crop diseases with resistant varieties. They shop aggressively for disease resistance, but they're becoming discouraged. No resistance is currently available to prevent the beetle-transmitted seed staining that sometimes sends their soybeans to feed markets rather than to Japanese tofu buyers. "There's very little public plant breeding going on right now," says Rosmann. "The interest is in biotechnology and that's where the dollars are going, sad to say."

His ridge-tilled fields are much cleaner than conventionally tilled fields, with only one-seventh to one-tenth as many weeds. Early tillage, rotary hoeing after planting and cultivation destroy most of the weeds in Rosmann's other fields. The rest of his weeds he simply lives with, peaceably and profitably.

Abundant small fields foster diverse practices

Rosmann Family Farms has several advantages many other farms don't: although they used pesticides for about 10 years during the 1960s and 1970s, the family never abandoned its mixed crop-livestock approach nor its generous crop rotations. In addition, the Rosmanns' 600-plus acres give them exceptional flexibility — and protection. "We have such a diversity of fields in different locations that we generally don't have problems in all of our fields at once — just a portion of a field."

The Rosmanns' practices are as diverse as their crops. They rotate some of their crop fields into grass-legume pastures, especially if those fields are building up unacceptable levels of weeds. They use cover crops in the corn they plant for silage but not in other corn fields. They rotate their grazing as well as their crops, thereby improving their pasture productivity and pest control. To provide feed for their cattle from mid-September until late fall, when corn stalks become available, they also follow barley and oats with turnips, rye and hairy vetch in mid-July.

The Rosmanns have been evaluating their individual practices with on-farm research trials for 15 years. They know what contributes to yield improvements and what doesn't but they haven't precisely pinpointed cause and effect — or whether interactions, rather than discrete practices, produce crop and soil benefits.

"There's no doubt, absolutely no doubt, that our approach is better for the environment and for us," Rosmann says. "But we just plain need research — on-farm systems research — to answer questions on farms like ours."

3 Principles of Ecologically Based Pest Management

 A WHOLE FARM APPROACH calls for designing a system that integrates ecological pest management into other aspects of crop and soil management. Each decision you make in designing your system for managing pests should be based in part on the impacts on the rest of the system.

Your steps toward implementing ecological pest management should be linked with soil organic matter management, soil nutrient management, tillage, and other efforts to reduce erosion and compaction. Creating field boundaries, borders and buffers designed to protect waterways also can lead to positive impacts on pest populations.

The following sections outline management strategies designed to augment the good bugs that will help ward off pests. You will learn ways to select plants that attract and feed beneficial insects, manage habitat to discourage pests, exploit plant breeding and natural plant defenses in your system, and maintain and improve soil diversity to benefit plant health.

Managing Aboveground Habitat

Diversify plants within agroecosystems. You can attract natural enemies and improve biological pest control by planting polycultures of annual crops — two or more crops simultaneously growing in close proximity. You can also let some flowering weeds reach tolerable levels or use cover crops such as buckwheat or sunflowers under orchards and vineyards.

For three decades, Dick Thompson has planted cover crops, managed weeds like covers instead of like pests, and lengthened and expanded his crop rotation. "I'm not saying we don't have any insect problems, but they do not constitute a crisis," says Thompson, who farms in Boone, Iowa. "We don't have to treat for them. We haven't done that for years."

Numerous researchers have shown that increasing plant diversity — and thereby habitat diversity — favors the abundance and effectiveness of natural enemies:

- In the Latin American tropics, lower numbers of leafhoppers and leaf beetles have been reported in small farms where beans are intercropped with corn. Corn earworm populations were reduced when corn was intercropped with legumes.
- In Canadian apple orchards, 4 to 18 times as many pests were parasitized when wildflowers were numerous compared to when they were few. In this research, wild parsnip, wild carrot and buttercup proved essential to maintaining populations of a number of parasitoids.
- In California organic vineyards, growing buckwheat and sunflowers between the vines attracts general predators as well as the leafhopper egg wasp (*Anagrus* species) to help manage grape leafhoppers and

Gary Kramer, USDA NRCS

Cover crops in a California orchard reduce soil erosion and contribute to overall farm diversity.

thrips. When these summer-blooming cover crops flower early, they allow populations of beneficials to surge ahead of pests. When they keep flowering throughout the growing season, they provide constant supplies of pollen, nectar and alternative prey. Mowing every other row of cover crops is a management practice that forces those beneficials out of the resource-rich cover crops and into vines.

- Georgia cotton fields strip-cropped with alfalfa or sorghum had higher populations of natural enemies that attack moth and butterfly pests. Beneficials reduced pest insects below economic threshold levels in cotton that was relay-cropped with crimson clover, eliminating the need for insecticides.

- At Michigan State University, researchers discouraged potato leafhoppers in alfalfa by adding forage grasses to alfalfa stands. The grasses don't provide the leafhoppers with enough nutrition to develop eggs, but the leafhoppers feed on them anyway for 5 to 8 minutes before trying another plant and eventually flying away. By diverting leafhoppers from alfalfa and by increasing their chances for dispersal, alfalfa-orchardgrass mixtures held 30 percent fewer leafhoppers than pure alfalfa stands. Because potato leafhoppers are often controlled later in the season by a naturally occurring fungus, this strategy may reduce leafhopper damage below threshold levels.

Zach Berkowitz

A mixture of perennial rye and chewings fescue helps moderate vigorous vine growth in deep valley soils. Grasses go dormant in the summer and begin growing again in the fall. See page 30.

Strategies to Enhance Beneficials

One of the most powerful and long-lasting ways to minimize economic damage from pests is to boost populations of existing or naturally occurring beneficial organisms by supplying them with appropriate habitat and alternative food sources. Beneficial organisms such as predators, parasites and pest-sickening "pathogens" are found far more frequently on diverse farms where fewer pesticides are used, than in monocultures or in fields routinely treated with pesticides.

The following characteristics are typical of farms that host plentiful populations of beneficials:

- Fields are small and surrounded by natural vegetation.
- Cropping systems are diverse and plant populations in or around fields include perennials and flowering plants.
- Crops are managed organically or with minimal agrichemicals.
- Soils are high in organic matter and biological activity and — during the off-season — covered with mulch or vegetation.

To conserve and develop rich populations of natural enemies, avoid cropping practices that harm beneficials. Instead, substitute methods that enhance their survival. Start by reversing practices that disrupt natural biological control, such as insecticide applications, hedge removal and comprehensive herbicide use intended to eliminate weeds in and around fields.

Even small changes in farming routines can substantially increase natural enemy populations during critical periods of the growing season. The simple use of straw mulch provides humid, sheltered hiding places for nocturnal predators like spiders and ground beetles. By decreasing the visual contrast between foliage and bare soil, straw mulch also can make it harder for flying pests like aphids and leafhoppers to "see" the crops they attack. This combination of effects can significantly reduce insect damage in mulched garden plots.

INNOVATIVE TART CHERRY ORCHARD SYSTEMS

Michigan State University scientists have evaluated orchard-scale ground cover experiments in established commercial orchards and in a new tart cherry orchard at the Northwest Horticultural Research Station. They studied orchard floors covered with compost, mulch or cover crops such as crimson clover, berseem clover, white clover, white Dutch clover, Michigan red clover, crown vetch, indigo vetch, alfalfa, rye, annual ryegrass, hard fescue and Buffalo grass. So far, findings include:

- Season-long populations of beneficial mites were attributed to the use of a red clover cover crop.
- Season-long, vegetation-free strips using either herbicide or mulch increase pest mite populations.
- Orchards with ground covers—irrigated but not treated with herbicides to manage weeds—had fruit yields that were not significantly lower than conventional practices over a five-year period. Note the irrigation may be critical in this system to prevent the ground cover from competing with the fruit trees for water.
- Adding mulch, cover crops and/or compost increases soil organic matter, populations of beneficial soil microbes and amounts of active soil carbon and nitrogen available to trees.
- Fewer beneficial nematodes, more plant-parasitic nematodes and more nitrate leaching were associated with lower-quality conventional-system soils.
- Hay or straw mulch, applied 6 to 8 inches deep, improved tree growth and yields despite higher pest mite populations.
- Nitrate leaching—greatest in spring and fall—was substantially reduced by vegetation growing under trees during these periods.

- In-row soil population densities of beneficial nematodes, mycorrhizae and earthworms were greater under an organic production system.
- Young trees benefited from adding mulch or compost but can be severely stunted by competition with groundcover plants for moisture and nutrients.
- Trees with heavy mulches produced soft fruit in two of seven years.

The scientists also are examining the impact of mixed-species hedgerows on insect pest movement into and out of orchards. In addition, they are evaluating insect pheromone mating disruption, mass trapping of plum curculio,14-inch groundcover bands around mulched center lines, and intercropping with such income-generating woody species as sea buckthorn and Siberian pea.

Charles Edson, Mich. State Univ.

Season-long populations of beneficial mites were attributed to the use of a red clover cover crop.

Orchards offer advantages over annual row crops in biological pest control, says MSU IPM tree fruit integrator David Epstein. Because they do not undergo major renovation every year, orchard systems can be developed to let beneficials get established. "Ground covers can be used to encourage beneficials to build up their populations and remain in the orchard throughout the year," he says.

How much the beneficials actually reduce pests, however, depends on weather, pest populations and the effectiveness of growers' monitoring programs. "To say that if you plant red clover you'll never have to spray for mites again would be erroneous," says Epstein. "But if you know what's out there—what levels of pests, predators and parasitoids you have—then you can make an informed decision as to whether or not you can save a spray." (For more information about this project, see http://www.ipm.msu.edu/tartcherry.htm)

As with most strategies described in this book, multiple benefits accrue from diversification. For example, carefully selected flowering plants or trees in field margins can be important sources of beneficial insects, but they also can modify crop microclimate, add organic matter and produce wood or forage. Establishing wild flower margins around crop fields enhances the abundance of beneficial insects searching for pollen and nectar. The beneficials then move into adjacent fields to help regulate insect pests. As an added benefit, many of these flowers are excellent food for bees, enhancing honey production, or they can be sold as cut flowers, improving farm income.

Straw mulch provides hiding places for such nocturnal predators as spiders and ground beetles.

CAUTION! Using mulch to increase populations of spiders and ground beetles only works if the pests attacking your crops are prey for those predators.

Increase the population of natural enemies. To an insect pest, a fertilized, weeded and watered monoculture is a dense, pure concentration of its favorite food. Many have adapted to these simple cropping systems over time. Natural enemies, however, do not fare as well because they are adapted to natural systems. Tilling, weeding, spraying, harvesting and other typical farming activities damage habitat for beneficials. Try instead to support their biological needs.

To complete their life cycles, natural enemies need more than prey and hosts; they also need refuge sites and alternative food. For example, many adult parasites sustain themselves with pollen and nectar from nearby flowering weeds while searching for hosts. Predaceous ground beetles — like many other natural enemies — do not disperse far from their overwintering sites; access to permanent habitat near or within the field gives them a jump-start on early pest populations.

Provide supplementary resources. You can enhance populations of natural enemies by providing resources to attract or keep them on your farm. In North Carolina, for example, erecting artificial nesting structures for the red wasp (*Polistes annularis*) intensified its predation of cotton leafworms and tobacco hornworms. In California alfalfa and cotton plots, providing mixtures of hydrolyzate, sugar and water increased egg-laying by green lacewings six-fold and boosted populations of predatory syrphid flies, lady beetles and soft-winged flower beetles.

You can increase the survival and reproduction of beneficial insects by allowing permanent populations of alternative prey to fluctuate below damaging levels. Use plants that host alternative prey to achieve this; plant them around your fields or even as strips within your fields. In cabbage, the relative abundance of aphids helps determine the effectiveness of the general predators that consume diamondback moth larvae. Similarly, in many regions, anthocorid bugs benefit from alternative prey when their preferred prey, western flower thrips, are scarce.

Select flowering plants that attract beneficial insects, such as this adult syrphid fly.

Another strategy is to augment the population of a beneficial insect's preferred host. For example, cabbage butterflies (a pest of cole crops) are the preferred host for two parasites (*Trichogramma evanescens* and *Apanteles rebecula*). Supplemented with continual releases of fertile females, populations of this pest escalated nearly ten-fold in spring. This enabled populations of the two parasites — both parasitic wasps — to buildup earlier in spring and maintain themselves at effective levels all season long. Because of its obvious risks, this strategy should be restricted to situations where sources of pollen, nectar or alternative prey simply can't be obtained.

Manage vegetation in field margins. With careful planning, you can turn your field margins into reservoirs of natural enemies. These habitats can be important overwintering sites for the predators of crop pests. They also can provide natural enemies with pollen, nectar and other resources.

Many studies have shown that beneficial arthropods do indeed move into crops from field margins, and biological control is usually more effective in crop rows near wild vegetation than in field centers:

- In Germany, parasitism of the rape pollen beetle is about 50 percent greater at the edges of fields than in the middle.
- In Michigan, European corn borers at the outskirts of fields are more prone to parasitism by the ichneumonid wasp *Eriborus terebrans*.
- In Hawaiian sugar cane, nectar-bearing plants in field margins improve the numbers and efficiency of the sugar cane weevil parasite (*Lixophaga sphenophori*).
- In California, where the egg parasite *Anagrus epos* (a parasitic wasp) reduces grape leafhopper populations in vineyards adjacent to French prunes, the prunes harbor an economically insignificant leafhopper whose eggs provide Anagrus with its only winter food and shelter.

In spring, lady beetles, which can live up to three years, feed on pollen and nectar from flowers, weeds and trees.

Lady beetles migrate into crop or vegetable fields, where they lay eggs near prey.

Eggs hatch and progress through several larval stages.

In fall and winter, lady beetles cluster under leaf litter, rocks, bark or other sheltered places.

In summer, lady beetle adults and larvae inhabit crop fields, preying primarily on aphids.

Lady beetles follow food sources from field margins into cash crops over the course of the season.

FARM FEATURE

NO-TILL COVER CROPS YIELD SOIL AND PEST BENEFITS

- Uses conservation tillage, manure and cover crops to manage pests
- Uses cover crops to conserve moisture
- Integrates crop and livestock operations

With slopes as steep as 7 percent and winds that sandblast his seedlings, Mark Vickers decided to try no-till production and cover crops on his Coffee County, Ga., farm nine years ago. A fourth-generation cotton and peanut grower who also plants corn or soybeans when the market is right, Vickers assumed his conservation-tillage system would keep his highly erodible soils in place.

It did that, but it also did a whole lot more. Along with regular manuring with poultry litter, Vickers' new farming practices eased many of his pest problems. Moreover, it made a "night and day" difference in his soils. "There's just no comparison," he says. "It's beginning to resemble potting soil rather than clay."

Production costs decrease by up to a third

With the cover crop acting much like "a jacket," Vickers' healthier soils hold moisture, prevent runoff and stretch his irrigation dollars. In its entirety, his farming system trims a quarter to a third off Vickers' production costs — mostly for labor, equipment and fuel. He sidedresses a bit of nitrogen and applies several conventional herbicides, but cutting back to just one preplant insecticide in his peanuts slashed the insecticide share of his budget by 50 to 60 percent.

Vickers now plants *Bt* cotton against bollworms and hasn't used insecticides against any cotton pests for the past two years. Ladybugs, fire ants, wasps, assassin bugs and bigeyed bugs are abundant in his fields. "It took between three and four years to build up the beneficial populations," he says. "I still have the same pests, but the beneficials seem to be keeping them in check and not letting them get over the threshold numbers."

Historically, Vickers has rarely been plagued with insects in his peanuts. When corn earworms uncharacteristically erupted last year, he treated them with pyrethroids. On the other hand, infestations of white mold and tomato spotted wilt virus were common occurrences before Vickers began using cover crops. He hasn't seen either of those diseases in his peanuts since.

Standout cover crop is rye

Although Vickers grows wheat, rye and oats as high-residue winter covers — and also sells the oats — it's the rye that's made him a believer in the value of cover crops. He uses it to prevent root-knot nematode problems and credits it with "dramatically" boosting his weed control, deterring weeds and "shading everything out."

Vickers sows his cover crops all the way to his field edges and even into his roads. He feeds them lightly with nitrogen if he thinks they need it. In spring, when he plants his summer cash crop, he kills the cover crop with a herbicide and plants either peanuts or cotton right into the standing litter. When he grows corn, he sows that directly into the green cover crop.

Cutting back to just one preplant insecticide in his peanuts slashed the insecticide share of his budget by 50 to 60 percent.

Vickers' improved farming practices let him produce profitable cash crops without hiring labor. "I do all of it myself — everything — but there's plenty of time to do it," he says. "If I weren't doing it this way, I couldn't farm. There would not be enough time for me to do everything that needed to be done to plant a crop."

Minimum-till: From "no way" to a better way

Mike Nugent, another Coffee County cotton and oat grower, says his minimum-till system has increased his cotton yields by half — to about 1,250 pounds of cotton lint per acre. Nugent plants an oat cover crop in late fall, lets his cattle graze it for a few months in winter and still harvests 80 bushels of certified oat seed per acre in spring. He irrigates about 40 percent less than the county's conventional farmers and saves $30 to $40 an acre on insecticides, not including application costs.

"If you had told me 10 years ago that I would be farming no-till, I would have laughed at you and said there was no way in the world that would work," says Nugent.

However, when he began easing into his new system seven years ago, he was struck by how many more beneficials inhabited his cover crop than his still-conventional fields. It's been three years now since Nugent last sprayed his cotton for budworms or bollworms. He uses herbicides in his Roundup Ready® cotton and treats his seed with fungicides, but he relies on scouting to manage his insect pests.

"You have to watch what you're doing," he says. "If they ever get out of hand, we'll have to spray them. But we let the populations get to a certain point, because the beneficials won't stay without anything to eat." Even when pest populations reach threshold levels, Nugent keeps scouting for another few days. "I've seen lots of times, when you wait one day and scout again, the population comes down — and once it comes down, it will keep coming down every day."

Less tillage, fewer weeds

The very first year he strip-tilled, Nugent also noticed many more weeds where he tilled than where he didn't. He responded by narrowing the tilled strip and now needs a third less herbicide.

He also uses very little commercial fertilizer, depending instead on poultry litter and on recycled nutrients that are pulled from the ground by his oats and returned through their dead straw.

In the beginning, his neighbors were skeptical. Now, Nugent says, "they've switched over, too, because they've seen that it works. It's the only way to farm, as far as I'm concerned."

FARM FEATURE

A TOAST TO ECOLOGICAL GRAPE PRODUCTION

- Uses cover crops to enhance beneficials and restrain plant vigor
- Manages riparian vegetation to reduce pests
- Matches flowers to resource needs of beneficials

Few wine drinkers are in the market for Cabernet Sauvignon with hints of asparagus or green pepper — herbaceous or "green" characters prompted by overly vigorous vines. Fewer still want utterly tasteless wines that have been drained of their flavors by spider mites.

In the vineyards of California's North Coast, consultant Zach Berkowitz's clients know that their wines will inevitably tell the tale of how their grapes were grown. During his three decades of advising grape growers, Berkowitz has learned that some pest management methods favor flavor while others put it at risk.

Berkowitz, who calls himself a "first-generation farmer," earned a degree in plant science from the University of California-Davis in 1980. Long committed to sustainable production, he says what he learned there about integrated pest management "immediately struck a chord." Now working with 10 or more growers and 1,500 or more acres — mostly in Napa and Sonoma counties — he tries to encourage beneficial organisms to keep production systems in balance while he manages for superior wine quality.

Start with cover crops

At the very least, Berkowitz says all grape growers can sow a no-till cover crop in the highly trafficked "avenues" surrounding their vineyards. "If those areas are seeded and mowed, that helps keep down dust, which helps keep down mites."

He also advises his clients to plant either annual or perennial cover crops in their vineyard rows — preferably between mid-September and mid-October. For vineyards whose soil is shallow or whose vines aren't strong, he recommends an annual mix of 'Zorro' fescue, 'Blando' brome and clovers. For those on flatter ground and with stronger vines, he prefers blends of such native perennial grasses as California brome, meadow barley and blue wildrye.

By curbing the vines' excessive vigor, these cover crops boost the grapes' appeal to wine drinkers and diminish their palatability to western grape leafhoppers. Berkowitz suspects that cover crops — especially "insectary" blends of flowering plants — also intensify populations of spiders, lacewings and other natural enemies of leafhoppers, thrips and mites.

During decades of advising grape growers, Berkowitz has learned that some pest management methods favor flavor while others put it at risk.

"It's kind of a subtle effect, but I think that over time the advantage increases," says Berkowitz. "You get that natural balance happening and it seems like your pest problems decrease."

Densely forested creeks surround many North Coast vineyards. "We're not cultivating fenceline to fenceline; we're striving to avoid monoculture," Berkowitz says. There's "reason to believe" this additional biodiversity contributes to pest control, he says, but more research would help.

Patience pays

Berkowitz says he likes to "preach patience," especially in managing fall-planted annual cover-crop mixes. "People want to mow it so it looks nice and tidy, but it's best to just let it go to seed," he says. By delaying mowing until April or May, growers can watch their thick layer of thatch turn golden brown in summer, then germinate naturally with the fall rains.

He makes an exception if the annual cover crop is infested with tall-growing mustards or other "junky resident weeds." Then, growers should mow first in January or February before those weeds set seed, setting their blades high enough to safely clear the crop. Repeated over several years, this process eventually creates the right conditions for the cover crop to dominate and the weeds to "kind of go away."

By late spring, when his clients mow their perennial grass-legume mixes, those cover crops have also served as alternate hosts for natural enemies. Berkowitz's experience indicates that, in the long term, even grass cover crops trim populations of leafhoppers, though not necessarily below economic thresholds.

Sow and mow strategically

Some growers like to cultivate every other row of their cover crops in early April and mow the rows in between. Then, they disk the mowed rows in May. To Berkowitz, that's better for pest management than mowing too early and almost as good as allowing the covers to go entirely to seed.

Other growers — not ready for a solid floor of no-till cover crops — don't plant those alternate rows to begin with. Instead, they simply sow every other row. Berkowitz endorses that practice for sites where soils aren't rich or deep and vines aren't overly vigorous. "It gives producers a little bit of a compromise and over time they can go to complete no-till."

Berkowitz cautions growers not to overfertilize insectary blends, whose energy should go towards flowering rather than towards vegetative growth, and he advises against fertilizing grass-legume mixtures at all, since the legumes will eventually help supply nitrogen. He supports fertilizing solely when growers of grass-only covers want "quick, thick" stands for erosion control.

Manage flexibly and responsively

On the rare occasions that leafhoppers or thrips exceed economic thresholds in his clients' vineyards, Berkowitz recommends insecticides. "We try to use systems that control pests without chemicals, but sometimes you're just stuck."

That's often the case with Pierce's disease, whose damage can force frequent replanting. Berkowitz says insecticide treatment for the blue-green sharpshooter during the first hot spells can regulate this vector's early movement into vineyards. Another approach showing "some merit" is riparian vegetation management: replacing host plants with non-hosts. This reduces the sharpshooter's populations while broadening diversity. "Today we try to manage the vector, but someday we hope to be able to control the disease itself," he says.

California vintners seed mixtures of Blando brome grass, Zorro fescue and crimson and rose clovers to prevent erosion, regulate vine growth and attract beneficial insects.

Over the years, Berkowitz has learned not to include 'Berber' orchard grass or annual ryegrass in cover crop mixtures because they're simply too competitive with grapevines. He has also observed that using sulfur to organically control powdery mildew kills predaceous beneficial mites faster than its kills prey mites.

"You think you're doing a good thing by dusting with sulfur, but at the end of the season you wind up with these mite problems." In vineyards where this has occurred, Berkowitz advises producers to substitute non-sulfur controls like biofungicides after early-spring treatments with sulfur. He has watched that strategy "really help" in repeatedly mite-infested vineyards.

"It's a systems approach," says Berkowitz. "That's what makes sustainable agriculture interesting to me: everything is connected."

Conservation filter strips can include flowering plants to attract beneficials and provide quality habitat for many species of wildlife.

In Norway's apple orchards, the abundance of apple fruit moth pests depends largely on the amount of berries produced by the European mountain ash (*Sorbus aucuparia*), a wild shrub. Because only one apple fruit moth larva can develop inside each berry, the number of these pests is directly limited by the number of berries. Thus, when European mountain ash fails to bear, apple fruit moth populations fail as well. Unfortunately, that also spells death for a naturally occurring parasite of the apple fruit moth, the braconid parasitoid wasp (*Microgaster politus*). Entomologists have advised Norwegian orchardists to plant a cultivated *Sorbus* (ash) for its regular and abundant crops. By sustaining both apple fruit moths and *Microgaster*, this practice allows the natural enemies to hold the moths at levels *Sorbus* can support. The result: the moths don't abandon *Sorbus* for orchards.

Manage plants surrounding fields to manage specific pests. One practice, called perimeter trap cropping, works best when plants like snap beans or cowpeas are grown to attract stink bugs and Mexican bean beetles away from soybeans. In perimeter trap cropping, plants that are especially attractive to target pests are planted around a cash crop, encircling it completely without gaps.

REDUCE MOWING FREQUENCY TO INCREASE BENEFICIALS

Tree fruit growers seeking alternatives to broad-spectrum pesticides are looking to manage insect pests using a more environmentally friendly approach. In Washington state pear orchards, SARE-funded research has found that mowing once a month rather than two or three times a month creates alluring habitats for beneficial insects.

An ARS researcher partly funded by SARE ran trials at three orchards and varied mowing frequency (weekly, monthly and just once a season). With less frequent mowing, the natural enemies moved into the ground cover in greater numbers, likely attracted to the pollen and nectar newly available from flowering plants as well as more abundant prey, such as aphids and thrips. Researcher Dave Horton found more lacewing larvae, spiders, ladybug beetles, damsel bugs, parasitoids and minute pirate bugs. "If you mow a lot, you won't have much in the way of natural enemies on the ground," Horton said. "By reducing the frequency to once a month, you see a dramatic increase in natural enemies moving into the ground cover without a big increase in pests that feed on fruit."

Questions remain whether the predators migrate from the ground cover into the pear trees to attack orchard pests, although evidence supports that some predators, especially spiders, appeared in higher numbers in pear trees in the less frequently mowed plots, good news for pear growers.

One of Horton's farmer collaborators, who received a SARE farmer/rancher grant to study similar ways to manage orchard pests, is convinced that minimal mowing provides control. "I'm practicing this, and I've never had to spray for mites," said George Ing of Hood River, Ore., who has a 16-year-old orchard. "Other orchards that are conventionally treated have more pests. I'm convinced it helped." At the behest of area growers, who provided a research grant through their pear

USDA ARS

and apple association, Horton will test how seeding cover crops such as white clover between tree rows affects populations of both pests and pest predators.

USDA ARS researcher Dave Horton found that less frequent mowing in orchards attracts more beneficial insects to prey on pear psylla, leaf miners and other serious pests.

Perimeter trap cropping can sharply reduce pesticide applications by attracting pests away from the cash crop. By limiting pesticide use in field borders or eliminating it entirely, you can preserve the beneficials in the main crop. Extension vegetable educators at the University of Connecticut report that up to 92 percent of pepper maggot infestation occurs on trap crops of unsprayed hot cherry peppers, effectively protecting the sweet bell peppers inside. Applying pesticide to the trap crop during the flight of the adult pepper maggot fly reduces infestations in the unsprayed bell peppers by 98 to 100 percent. Connecticut commercial growers with low to moderate pepper maggot populations have confirmed the method's success on fields as small as one-quarter acre and as large as 20 acres.

In Florida, researchers with the USDA-ARS found that a collard trap crop barrier around commercial cabbage fields prevented diamondback moth larvae from exceeding action thresholds and

T. Jude Boucher, Univ. of Conn.

Planting squash as a "trap" crop draws pests like cucumber beetles away from cash crops, reducing insecticide use and improving yields.

acted as a refuge planting to build parasite numbers; cabbage growers who used perimeter trap cropping reduced their insecticide applications by 56 percent. In Ontario, Canada, researchers also found that planting 'Southern Giant' mustard around fields of cabbage, cauliflower and broccoli protected them from flea beetles.

Alternately, flowering plants such as *Phacelia* or buckwheat can be grown in field margins to increase populations of syrphid flies and reduce aphid populations in adjacent vegetable crops. This method is most effective for pests of intermediate mobility. Consider plants that support beneficial insects and can be harvested to generate revenue.

Create corridors for natural enemies. You can provide natural enemies with highways of habitat by sowing diverse flowering plants into strips every 165 to 330 feet (50–100 m) across the field. Beneficials can use these corridors to circulate and disperse into field centers.

European studies have confirmed that this practice increases the diversity and abundance of natural enemies. When sugar beet fields were drilled with corridors of tansy leaf *(Phacelia tanacetifolia)* every 20 to 30 rows, destruction of bean aphids by syrphids intensified. Similarly, strips of buckwheat and tansy leaf in Swiss cabbage fields increased populations of a small parasitic wasp that attacks the cabbage aphid. Because of its long summer flowering period, tansy leaf has also been used as a pollen source to boost syrphid populations in cereal fields.

> TIP Sowing diverse flowering plants, such as tansy leaf and buckwheat, into strips that cut across fields every 165 to 330 feet (50–100 m) can provide natural enemies with highways of habitat.

For more extended effects, plant corridors with longer-flowering shrubs. In northern California, researchers connected a riparian forest with the center of a large monoculture vineyard using a vegetational corridor of 60 plant species. This corridor, which included many woody and herbaceous

Miguel Altieri, Univ. of Calif.

A corridor of Alyssum acts as a highway of habitat drawing beneficial insects into this large field of lettuce.

perennials, bloomed throughout the growing season, furnishing natural enemies with a constant supply of alternative foods and breaking their strict dependence on grape-eating pests. A complex of predators entered the vineyard sooner, circulating continuously and thoroughly through the vines. The subsequent food-chain interactions enriched populations of natural enemies and curbed numbers of leafhoppers and thrips. These impacts were measured on vines as far as 100 to 150 feet (30–45 m) from the corridor.

BEETLE BANKS BOOST BENEFICIALS

Some grass species can be important for natural enemies. For example, they can provide temperature-moderating overwintering habitats for predaceous ground beetles. In England, researchers established "beetle banks" by sowing earth ridges with orchard grass at the centers of cereal fields. Recreating the qualities of field boundaries that favor high densities of overwintering predators, these banks particularly boosted populations of two ground beetles (Demetrias atricapillus and Tachyporus hypnorum), important cereal aphid predators. A 1994 study found that the natural enemies the beetle banks harbored were so cost-effective in preventing cereal aphid outbreaks that pesticide savings outweighed the labor and seed costs required to establish them. The ridges in this study were 1.3 feet high, 5 feet wide and 950 feet long (0.4 m x 1.5 m x 290 m).

For more information, see "Habitat management to conserve natural enemies of arthropod pests in agriculture" (Resources, p. 104).

Jack Kelly Clark, Univ. of Calif.

Predaceous ground beetles feed mainly on caterpillars and other insect larvae.

Select the most appropriate plants. Beneficial insects are attracted to specific plants, so if you are trying to manage a specific pest, choose flowering plants that will attract the right beneficial insect(s). The size and shape of the blossoms dictate which insects will be able to access the flowers' pollen and nectar. For most beneficials, including parasitic wasps,

Rob Myers, Jefferson Institute

Buckwheat
(Fagopyrum esculentum)

Peggy Greb, USDA ARS

A cover crop of mustard can be disked into soil as "green manure" to act as a natural fumigant for weeds and diseases.

the most helpful blossoms are small and relatively open. Plants from the aster, carrot and buckwheat families are especially useful (Table 1).

Timing is as important to natural enemies as blossom size and shape, so also note when the flower produces pollen and nectar. Many beneficial insects are active only as adults and only for discrete periods during the growing season: They need pollen and nectar during those active times, particularly in the early season when prey is scarce. One of the easiest ways you can help is to provide mixtures of plants with relatively long, overlapping bloom times. Examples of flowering plant mixes might include species from the daisy or sunflower family (Compositae) and from the carrot family (Umbelliferae).

Information about which plants are the most useful sources of pollen, nectar, habitat and other critical needs is far from complete. Clearly, many plants encourage natural enemies, but scientists have much more to learn about which plants are associated with which beneficials and how and when to make desirable plants available to key predators and parasitoids. Because

SURROUNDING CROPS WITH PERIMETER FOOLS PESTS

Nelson Cecarelli of Northford, Conn., who often lost an entire season's cucumber crop to voracious cucumber beetles, planted squash around his field perimeter, sprayed minimally, and harvested a bounty of cukes in 2003 and 2004. Cecarelli was one of about 30 farmers in New England to adopt a perimeter trap cropping strategy recommended by a University of Connecticut researcher who, with a SARE grant, tested the theory over two seasons – with terrific results. The system, popular among growers, encircles a vulnerable vegetable with a crop that can attract and better withstand pest pressure, reducing the need for pesticides.

"What you're seeking in a trap crop is something that gets up and out of the ground fast with lots of foliage and won't be over-run easily when beetles come into the field," said T. Jude Boucher, Extension Educator and project leader, who recommends a thick-skinned squash called Blue Hubbard. "If we can stop beetles during the seedling stage, we can eliminate most of the damage."

In 2004, nine New England growers, including Randy Blackmer (below), increased yields of cucumbers and summer squash by 18 percent and reduced insecticide use by 96 percent, earning an extra $11,000 each, on average, Boucher said. The research compared a dozen farms using perimeter trap cropping to farms that used the typical regimen of four sprays per year.

Growers planting perimeters applauded the time savings in pest scouting and pesticide spraying, and the improved economics thanks to lower input costs and higher, better-quality yields.

Despite pessimism that the Blue Hubbard squash wouldn't appeal to customers, most participating farmers found that Blue Hubbard resisted beetle damage and sold at their markets. In post-project surveys, farmers said the system not only saved money, but also that planting a perimeter was simpler than applying multiple full-field insecticide sprays.

"We're trying to get away from the 'silver bullet' mentality that you can put on a pesticide and it'll stop your problem," Boucher said. "We're changing the pest populations' dynamics in the field."

T. Jude Boucher, Univ. of Conn.

Randy Blackmer examines pumpkins planted as a trap crop to draw cucumber beetles away from squash on his Connecticut farm.

Sunflowers in California vineyards draw the leafhopper egg wasp, a parasite of the grape leafhopper.

beneficial interactions are site-specific, geographic location and overall farm management are critical variables. In lieu of universal recommendations, which are impossible to make, you can discover many answers for yourself by investigating the usefulness of alternative flowering plants on your farm. Also consider tapping into informational networks, such as Extension, NRCS and nonprofit organizations. Other farmers make great information sources, too (Resources, p. 104).

Use weeds to attract beneficials. Sometimes, the best flowering plant to attract beneficials is a weed, but this practice complicates management. Although some weeds support insect *pests*, harbor plant diseases or compete with the cash crop, others supply essential resources to many *beneficial* insects and contribute to the biodiversity of agroecosystems.

In the last 20 years, researchers have found that outbreaks of certain pests are less likely in weed-diversified cropping systems than in weed-free fields. In some cases, this is because weeds camouflage crops from colonizing pests, making the crops

TIP When using weeds in your biological control program, first define your pest management strategy precisely, then investigate the economic thresholds that weeds should not exceed.

TABLE 1
Flowering Plants That Attract Natural Enemies

COMMON NAME	GENUS AND SPECIES	PHOTO LOCATION
Umbelliferae (Carrot family)		
Caraway	*Carum carvi*	
Coriander (cilantro)	*Coriandrum sativum*	
Dill	*Anethum graveolens*	
Fennel	*Foeniculum vulgare*	
Flowering ammi or Bishop's flower	*Ammi majus*	
Queen Anne's lace (wild carrot)	*Daucus carota*	
Toothpick ammi	*Ammi visnaga*	
Wild parsnip	*Pastinaca sativa*	
Compositae (Aster family)		
Blanketflower	*Gaillardia* spp.	
Coneflower	*Echinacea* spp.	p. 52
Coreopsis	*Coreopsis* spp.	
Cosmos	*Cosmos* spp.	
Goldenrod	*Solidago* spp.	
Sunflower	*Helianthus* spp.	p. 41
Tansy	*Tanacetum vulgare*	
Yarrow	*Achillea* spp.	
Legumes		
Alfalfa	*Medicago sativa*	
Big flower vetch	*Vicia grandiflora*	
Fava bean	*Vicia fava*	
Hairy vetch	*Vicia villosa*	p. 75
Sweet clover	*Melilotus officinalis*	
Brassicaceae (Mustard family)		
Basket-of-gold alyssum	*Aurinium saxatilis*	
Hoary alyssum	*Berteroa incana*	
Mustards	*Brassica* spp.	p. 39
Sweet alyssum	*Lobularia maritima*	p. 37
Yellow rocket	*Barbarea vulgaris*	
Wild mustard	*Brassica kaber*	
Other species		
Buckwheat	*Fagopyrum esculentum*	p. 39
Cinquefoil	*Potentilla* spp.	

less apparent to their prospective attackers. In other cases, it is because the alternative resources provided by weeds support beneficials.

Unquestionably, weeds can stress crops, but substantial evidence suggests that farmers can enhance populations of beneficials by manipulating weed species and weed-management practices. A growing appreciation for the complex relationships among crops, weeds, pests and natural enemies is prompting many of today's farmers to emphasize weed *management* over weed *control.*

Dandelions are an important early-season source of nectar and pollen for beneficial insects.

Using weeds in your biological control program will require an investment of time and management skills. First, define your pest management strategy precisely, then investigate the economic thresholds that weeds should not exceed. If you choose to work with weeds in your bio-diverse farming system, consider the following management strategies:

- Space crops closely.
- Limit weeds to field margins, corridors, alternate rows or mowed clumps within fields, rather than letting them spread uniformly across fields.
- Use species sold in insectary plant mixtures.
- Prevent or minimize weed seed production.
- Mow weeds as needed to force beneficial insects into crops.
- Time soil disturbances carefully — for example, plow recently cropped fields during different seasons — so specific weeds can be available when specific beneficials need them.
- Except in organic systems, apply herbicides selectively to shift weed populations toward beneficial weed species.

Enhance plant defenses against pests. The first line of defense against insect pests is a healthy plant. Healthy plants are better able to withstand the onslaught of insect pests and can respond by mobilizing inbred mechanisms to fight off the attack. You can enhance natural defenses by improving soil and providing the best possible growing conditions, including adequate (but not excess) water and nutrients.

As plants co-evolved with pests, they developed numerous defenses against those pests. Some of those defenses have been strengthened over time through plant breeding, while others have been lost. Some plant defenses — spines, leaf hairs and tough, leathery leaves — are structural. Others are chemical:

- *Continuous,* or *constitutive* defenses are maintained at effective levels around the clock, regardless of the presence of pests; they include toxic plant chemicals that deter feeding, leaf waxes that form barriers, allelopathic chemicals that deter weed growth and other similar defenses.
- *Induced* responses are prompted by pest attacks; they allow plants to use their resources more flexibly, spending them on growth and reproduction when risks of infection or infestation are low but deploying them on an as-needed basis for defense when pests reach trigger levels.

The most effective and durable plant defense systems combine continuous and induced responses. Under attack by a plant-eating insect or mite, a crop may respond *directly* by unleashing a toxic chemical that will damage the pest or obstruct its feeding. It may also respond *indirectly,* recruiting the assistance of a third party.

Many plants produce volatile chemicals that attract the natural enemies of their attackers. To be effective, these signals must be identifiable and distinguishable by the predators and parasites whose help the

TIP Excess nitrogen fertilizer may hamper cotton's ability to send a chemical call for help.

crop is enlisting. Fortunately, plants under attack release different volatile compounds than plants that have not been damaged. Crops can even emit different blends of chemicals in response to feeding by different pests. Different varieties of the same plant — or even different parts or growth stages — can release different amounts and proportions of volatile compounds.

In response to attack by insect pests, cotton emits a chemical signal calling beneficial wasps to the rescue.

Leaves that escape injury also produce and release volatiles, intensifying the signaling capacities of damaged plants.

For example, when a beet armyworm chews on a cotton plant, the plant releases a specific chemical signal blend into the air. Female parasitic wasps pick up this signal and use it to locate the armyworm. They sting the armyworm and lay their eggs inside it, causing an immediate and dramatic reduction in armyworm feeding. This greatly reduces damage to the plant that originated the signal. Interestingly, inappropriate levels of added nitrogen can change the ratio of the molecules that comprise the chemical signal, thereby changing the signal and rendering it unnoticeable by the wasp.

CAUTION! Plant varieties are not equal in their abilities to defend themselves: some modern varieties lack the defenses of their native predecessors.

Plant breeding — though overwhelmingly beneficial in the short term — can have unforeseen consequences that unravel the best-laid plans of plant geneticists. Since the focus of plant breeding for pest resistance is often limited to a specific plant/pest interaction, selecting for one resistant gene could inadvertently eliminate other genes affecting other pests or genes that play a role in attracting the natural enemies of the pest.

In addition, newly developed varieties may stand better or yield more at the expense of natural defenses that are often unintentionally sacrificed for those other qualities. Selecting *for* one trait such as height could mean selecting inadvertently *against* any one of the many inborn plant defenses against pests. For example:

Horticulturist Philip Forsline examines hybrid grapes developed in a USDA breeding program.

- Scientists in Texas found that nectar-free cotton varieties attract fewer butterfly and moth pests, as their developers intended; however, as a consequence, these varieties also attract fewer parasites of tobacco budworm larvae and are thus more susceptible to that pest.

- According to USDA Agricultural Research Service scientists in Gainesville, Fla., today's higher-yielding commercial cottons produce volatile chemical signals at only one-seventh the level of naturalized varieties, impairing their ability to recruit natural enemies.

Fortunately, our knowledge of plants' roles in their own defense is steadily expanding. This knowledge can be used to breed and engineer plants whose defenses work harmoniously with natural systems. More research as well as plant breeding programs that focus on enhancing natural defenses are needed. Such programs might emphasize open-pollinated crops over hybrids for their adaptability to local environments and greater genetic diversity.

FARM FEATURE

RESISTANT FRUIT VARIETIES REDUCE RISK

- Suppresses annual weeds with mulches
- Improves soils with animal manure
- Uses disease-resistant varieties

Wisconsin fruit grower Eric Carlson pays twice the price of conventional fertilizers to feed his half-acre of transitional-organic blueberries with composted poultry manure, augmented with elemental sulfur, potassium and magnesium. He calculates that those blueberries need a half-mile of weeding every two or three weeks — a full mile if you figure both sides. The semi-load of mulches he buys each year suppresses his annual weeds, but perennials like sorrel and quackgrass — the latter so tenacious he's come to admire it — persist. At $8 an hour, Carlson's hand weeding costs five to 10 times as much as herbicide treatments.

"I know what I'm getting into, so I'm starting small," says Carlson. Fortunately, he has an urban customer base willing to pay what it costs to grow organic blueberries.

Because Carlson sells 95 percent of his fruit right on his Bayfield County farm — 70 percent of it pick-your-own — he also has customers eager to sample novel scab-resistant apples like *Jonafrees, Redfrees, Priscillas, Pristines* and *Liberties*. He doubts that would be case if he were selling his apples wholesale. Fortunately, his direct-market emphasis allows Carlson to take risks growing diverse varieties that other producers would be reluctant to try.

Carlson, who earned dual bachelor's degrees in horticulture and agronomy from the University of Wisconsin in 1983, first began following his dreams in 1989. That's when he left a seven-year job at the UW fruit pathology laboratory to grow his own hardy blueberries. Reared in the Milwaukee suburb of Wauwatosa, he chose 40 "exceptionally beautiful" acres on a finger of northern Wisconsin that juts into Lake Superior. Gradually, he expanded to 3 acres of blueberries, 1½ acres of raspberries, an acre of fresh-cut and everlasting flowers and 1,200 apple trees.

Economic sustainability comes first

Environmental sustainability has been an objective of Carlson's enterprise from the beginning. "We wanted people to come here, enjoy the environment and be able to walk around and buy healthy food," he says. However, financial reality quickly earned equal billing.

"You have to make the system economically sustainable first and then use the tools that are available to you to make it environmentally sustainable," says Carlson. "That's always been a struggle for me. My ideal is not using any synthetic chemicals, but I need to stay in business."

That's why Carlson now sparingly uses malathion to stop leafhoppers from infecting his flowers with aster yellows disease, which they can briskly do within 24 hours. With about a fifth of his 250 flower species susceptible to the plant-killing virus, Carlson scouts his fields daily when his climate is ripe for leafhoppers, spraying once or twice if he must.

His direct-market emphasis allows Carlson to take risks growing diverse varieties that other producers would be reluctant to try.

It's also why he has adopted a "low-spray" program for his apples, treating them conservatively with the relatively short-lived organophosphate Imidan: twice around petal-fall for plum curculio and codling moth and about twice after petal-fall for apple maggot flies. "I feel like it's the least amount that I can put out there and still have a marketable crop," he says. He times his apple maggot sprays with red visual traps.

Alternative disease management slashes fungicide use

For two years, Carlson cooperated with UW researchers as they built a predictive model for apple scab around measurements of air temperatures and leaf surface moisture. Some years, he uses only half as many fungicides as conventional growers do on his three scab-susceptible apple varieties — Cortland, Gala and Sweet 16 — while other years he can eliminate only one or two treatments.

On his 1,000 scab-resistant trees, which outnumber his susceptible trees five-fold, Carlson applies no fungicides at all. During the growing season, he quickly cuts out branches showing the earliest signs of fireblight and, during the dormant season, he aggressively prunes any possibly overwintering cankers. "Typically, apple growers spray tank mixtures of fungicides plus insecticides," says Carlson. "On my scab-resistant block, I'm not putting fungicides into the tank, so I feel good about that."

Carlson planted his apples densely — and consequently more expensively — on dwarfing rootstocks. That has allowed him to respond more nimbly to changing consumer tastes, since trees on dwarf rootstocks typically start bearing in two years rather than five. While his customers like learning that their apples were grown without fungicides, Carlson knows that flavor is what sells fruit and that consumer preferences can rival aroma compounds for volatility.

Rested raspberries reward their producers

Carlson is also experimenting with alternate-row production in raspberries. By mowing every other row of his berries, he hopes to significantly reduce fungicide applications and to use preemergence herbicides only once every three or four years. "You would think you would also cut your yields in half, but that's not necessarily the case," he says. "Because of how well the plant responds to a rest year, the research shows that you can get up to 75 percent of your normal production."

According to Carlson, a plethora of cane diseases make raspberries difficult to raise organically, so he grows them with what he calls a "basically conventional IPM approach." He trickle-irrigates them and makes sure 1½ to 2 feet of circulation-enhancing space separates his plants, minimizing the odds of raspberry disease.

After almost 15 years as an agricultural entrepreneur, Carlson likens fruit crops to "waves coming into shore." They don't produce harvests immediately but, like those waves, they "will come in the long run." Although working for himself — and for the health of his customers and the environment — is less predictable than his old university paycheck, Carlson makes sure he's still waiting on the shore by keeping his risks manageable.

Predators

COLEOPTERA: STAPHYLINIDAE	COLEOPTERA: CARABIDAE
Rove beetle *(Nudobius lentus)*	Pedunculate ground beetle *(Pasimachus depressus)*
COLEOPTERA: CARABIDAE	COLEOPTERA: MELYRIDAE
Ground beetle *(Calosoma sycophanta)*	Soft-winged flower beetle
COLEOPTERA: COCCINELLIDAE	COLEOPTERA: CANTHARIDAE
Sevenspotted lady beetle *(Coccinella septempunctata)*	Goldenrod Soldier Beetle *(Chauliognathus pennsylvanicus)*
DERMAPTERA: LABIDURIDAE	DIPTERA: SYRPHIDAE
Earwig	Syrphid fly

Predators

DIPTERA: SYRPHIDAE

Syrphid fly on aster

DIPTERA: SYRPHIDAE

Syrphid fly

HEMIPTERA: PENTATOMIDAE

Predatory rough shield stink bug
(*Brochymena* spp.)

HEMIPTERA: PENTATOMIDAE

Spined soldier bug
(*Podisus maculiventris*)

HEMIPTERA: PENTATOMIDAE

Predatory stink bug nymph

HEMIPTERA: NABIDAE

Damsel bug
(*Nabis alternatus*)

HEMIPTERA: LYGAEIDAE

Big-eyed bug with whitefly
(*Geocoris* spp.)

HEMIPTERA: REDUVIIDAE

Leafhopper assassin bug
(*Zelus renardii*)

Predators

HEMIPTERA: REDUVIIDAE	HEMIPTERA: REDUVIIDAE
	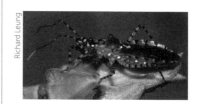
Assassin bug *(Apiomerus crassipes)*	**Striped Bug** *(Pselliopus cinctus)*
HEMIPTERA: ANTHOCORIDAE	HEMIPTERA: ANTHOCORIDAE

Adult minute pirate bug *(Orius tristicolor)*	**Pirate bug** *(Orius* spp.*)*
HEMIPTERA: ANTHOCORIDAE	ORTHOPTERA: MANTIDAE

Minute pirate bug nymph	**Praying mantid nymph on coneflower eating a fly**
NEUROPTERA: CHRYSOPIDAE	NEUROPTERA: CHRYSOPIDAE

Green lacewing larva *(Chrysoperla rufilabris)*	**Green lacewing adult** *(Chrysoperla carnea)*

Parasitoids

HYMENOPTERA: PTEROMALIDAE	HYMENOPTERA: ICHNEUMONIDAE
Scott Bauer, USDA ARS	Edward Holsten
Pteromalid ecto-parasitoid *(Catolaccus grandis)*	**Ichneumonid wasp** *(Glypta spp.)*
HYMENOPTERA: SCELIONIDAE	HYMENOPTERA: EULOPHIDAE
Richard Leung	Richard Leung
Scelionid wasp	**Eulophid wasp**
HYMENOPTERA: PTEROMALIDAE	HYMENOPTERA: ENCYRTIDAE
Richard Leung	Richard Leung
Chalcid wasp	**Encyrtid wasp**
HYMENOPTERA: BRACONIDAE	DIPTERA: TACHINIDAE
Scott Bauer, USDA ARS	Debbie Roos
Braconid wasp *(Peristenus digoneutis)*	**Tachinid fly**

Pests

HEMIPTERA: APHIDAE

Scott Bauer, USDA ARS

Green peach aphid
(Myzus persicae)

TETRANYCHIDAE: ACARI

Jack Kelly Clark, UCI

Two spotted spider mite
(Tetranychus urticae)

HEMIPTERA: PENTATOMIDAE

Russ Ottens

Green stink bug nymph
(Acrosternum hilare)

COLEOPTERA: CHRYSOMELIDAE

Scott Bauer, USDA ARS

Colorado potato beetle
(Leptinotarsa decemlineata)

HEMIPTERA: MIRIDAE

Scott Bauer, USDA ARS

Alfalfa plant bug
(Adelphocoris lineolatus)

HEMIPTERA: MIRIDAE

Scott Bauer, USDA ARS

Tarnished plant bug
(Lygus lineolaris (Palisot de Beauvois))

COLEOPTERA: CHRYSOMELIDAE

Scott Bauer, USDA ARS

Striped cucumber beetle
(Acalymma vittatu)

LEPIDOPTERA: NOCTUIDAE

Debbie Roos

Fall armyworm parasitized
by chalcid wasp

4 Managing Soils To Minimize Crop Pests

AGRICULTURAL PRACTICES THAT PROMOTE HEALTHY SOILS are a pillar of ecologically based pest management. Good soil management can improve water storage, drainage, nutrient availability and root development, all of which may, in turn, influence crop-defense mechanisms and populations of potential beneficials and pests.

In contrast, adverse soil conditions can hinder plants' abilities to use their natural defenses against insects, diseases, nematodes and weeds. Poor soils can cause plants to emit stress signals to potential attackers, heightening the risk of insect damage. For more information about improving your soil quality, see *Building Soils for Better Crops, 2nd Edition* (Resources, p. 104).

Healthy Soils Produce Healthy Crops

A healthy soil produces healthy crops with minimal amounts of external inputs and few to no adverse ecological effects. It contains favorable biological, physical and chemical properties.

A **biologically** healthy soil harbors a multitude of different organisms — microorganisms such as bacteria, fungi, amoebae and paramecia, as well as larger organisms like nematodes, springtails, insect larvae, ants, earthworms and ground beetles. Most are helpful to plants, enhancing the availability of nutrients and producing chemicals that stimulate plant growth.

Among the vital functions of soil organisms are:

- Breaking down litter and cycling nutrients
- Converting atmospheric nitrogen into organic forms and reconverting organic nitrogen into inorganic forms that plants can use
- Synthesizing enzymes, vitamins, hormones and other important substances
- Altering soil structure
- Eating and/or decomposing weed seeds
- Suppressing and/or feeding on soil-borne plant pathogens and plant-parasitic nematodes

A very compact soil has few large pores and thus is less hospitable to such organisms as springtails, mites and earthworms.

A healthy, biodiverse soil will support high levels of potentially beneficial soil organisms and low levels of potentially harmful ones. A soil rich in fresh residues — sometimes called particulate or light fraction organic matter — can feed huge numbers of organisms and foster abundant biological activity.

Dan Anderson, Univ. of Ill.

A soil's **physical** condition — its degree of compaction, capacity for water storage and ease of drainage — is also critical to soil and plant health. Good soil tilth promotes rainfall infiltration, thereby reducing runoff and allowing moisture to be stored for later plant use. It also encourages proper root development.

When aeration and water availability are ideal, plant health and growth benefit. For example, crops growing in friable soils with adequate aeration are less adversely affected by both wet

By adding cover crops and switching to no-till, Junior Upton drastically improved his habitually compacted soil.

and dry conditions than those growing in compacted soils. Soils with good physical structure remain sufficiently aerated during wet periods, and — in contrast to compacted soils — they are less likely to become physical barriers to root growth as conditions become very dry. Organic matter improves aeration by promoting the aggregation of soil particles. Secretions of mycorrhizal fungi, which flourish in organic matter, also improve a soil's physical properties.

Among the important **chemical** determinants of a soil's health are its pH, salt content and levels of available nutrients. Low quantities of nutrients, high levels of such toxic elements as aluminum and high concentrations of salts can adversely affect the growth of your crops. Healthy soils have adequate — but not excessive — nutrients. Excessive available nitrogen can make plants

David Nance, USDA ARS

Nitrogen-fixing symbiotic bacteria flourish in healthy soil.

more attractive or susceptible to insects, and overabundant nitrogen and phosphorus can pollute surface and groundwater. Well-decomposed organic matter helps healthy soils hold onto calcium, magnesium and potassium, keeping these nutrients in the plants' root zone.

The biological, physical and chemical aspects of soils all interact with and affect one another. For example, if your soil is very compact, it will have few large pores and thus will be less hospitable to such organisms as springtails, mites and earthworms. In addition, its lower levels of oxygen may influence both the forms of nutrients that are present and their availability; under anaerobic conditions, for instance, significant quantities of nitrate may be converted to gaseous nitrogen and lost to the atmosphere.

Managing Pests With Healthy Soils

Healthier soils produce crops that are less damaged by pests. Some soil-management practices boost plant-defense mechanisms, making plants more resistant and/or less attractive to pests. Other practices — or the favorable conditions they produce — restrict the severity of pest damage by decreasing pest numbers or building beneficials. Using multiple tactics — rather than one major tactic like a single pesticide — lessens pest damage through a third strategy: it diminishes the odds that a pest will adapt to the ecological pest management measures.

Practices that promote soil health constitute one of the fundamental pillars of ecological pest management. When stress is alleviated, a plant can better express its inherent abilities to resist pests (Figure 2). Ecological pest management emphasizes preventative strategies that enhance the "immunity" of the agroecosystem. Farmers should be cautious of using reactive management practices that may hinder the crop's immunity. Healthier soils also harbor more diverse and active populations of the soil organisms that compete with, antagonize and ultimately curb soil-borne pests. Some of those organisms — such as springtails — serve as alternate food for beneficials when pests are scarce, thus maintaining viable populations of beneficials in the field. You can favor beneficial organisms by using crop rotations, cover crops, animal manures and composts to supply them with additional food.

TIP Encourage beneficial organisms by using crop rotations, cover crops, animal manures and composts to supply them with additional food.

Potato plants grown in rye residue in plots run by USDA-ARS's Insect Biocontrol Lab (top) fare better than those grown using a system without cover crops (below).

In southern Georgia, cotton and peanut growers who planted rotation crops and annual high-residue winter cover crops, then virtually eliminated tillage, no longer have problems with thrips, bollworms, budworms, aphids, fall armyworms, beet armyworms and white flies. The farmers report that the insect pests declined after three years of rotations and cover crops. They now pay $50–$100 less per acre for more environmentally benign insect control materials such as *Bacillus thuringiensis* (Bt), pyrethroids and/or insect growth regulators.

In their no-till research plots with cover crops and long rotations, University of Georgia scientists haven't needed fungicides for nine years in peanuts, insecticides for 11 years in cotton, and insecticides, nematicides or fungicides for 17 years in vegetables. They also are helping growers of cucumbers, squash, peppers, eggplant, cabbage peanuts, soybeans and cotton reduce their pesticide applications to two or fewer while harvesting profitable crops. This system is described in greater detail in *Managing Cover Crops Profitably, 2nd Edition* (Resources, p. 104).

FARM FEATURE

TRIPLE THREAT TO PESTS: COVER CROPS, NO-TILL, ROTATION

- Uses cover crops to break up insect and disease cycles
- Releases parasites against pests
- Controls weeds with crop rotations, cover crops and no-till
- Uses no-till to conserve moisture

Since the early 1980s, Steve Groff has been building a sustainable farming system on the triple foundations of cover crops, intensive crop rotation and long-term no-tillage.

After more than 20 years — seven of them in no-till vegetables — Groff says he would "never come back" to conventional production. "I'm increasing beneficial insects to the degree that I'm getting a positive pest-control response. There's no doubt about that," he says. "But we haven't 'arrived' yet."

Crops need 40 percent less pesticide

Groff estimates that he has pared down pesticide use by 40 percent on his Cedar Meadow Farm in Lancaster County, Pa. By transplanting his 25 acres of tomatoes directly into rolled-down cover-crop mulch, he has sliced $125-an-acre from that crop's pesticide bill alone. His cover-crop mixes of hairy vetch, crimson clover and rye — or vetch and rye alone when clover is too expensive — harbor beneficials. They also seem to obstruct, exhaust, confuse and otherwise inhibit Colorado potato beetles, discouraging their colonization, says Aref Abdul-Baki, USDA Agricultural Research Service vegetable production specialist. Likewise, the killed cover crop may be dissuading cucumber beetles in Groff's 30 acres of pumpkins.

Groff says he hasn't sprayed his tomatoes against Colorado potato beetles for the past seven years, nor has he used post-emergence chemicals against cucumber beetles in pumpkins. He can also delay protective sprays for early blight for three to seven weeks in his tomatoes: in conventional systems, heavy raindrops pick up disease spores on plants, wash them down to plastic mulch, then splash them back up onto the crop; Groff's natural mulch lets spore-laden raindrops flow through to the ground, says Abdul-Baki. Similarly, the cover-crop mulch keeps his pumpkins cleaner and less prone to rot.

Integrated approach is essential

Although the mulches break up insect and disease cycles, Groff gives much of the pest-management credit to his long-term rotations. There's no single "magic bullet," he says: all three components of his system are equal partners.

In his 25 acres of sweet corn, Groff uses moth-trap monitoring to keep his corn earworm losses in check. In cooperation with a multi-state team of scientists led by Cornell University, Cedar Meadow Farm is also participating in investigatory releases of the parasitic wasp *Trichogramma ostriniae* against European corn borers.

For reasons he doesn't quite understand, Groff says aphids trouble none of his crops. He credits beneficials.

In exceptionally dry years, Groff's farm isn't spared from significant spider mite damage. "Right now we don't have a solution for that," he says. "This system is not foolproof." In wet years, he sees more slugs than his neighbors do. "Now, there is an instance where the residue and moisture definitely favor a pest," he says. Because gardeners' remedies like beer traps aren't even remotely economical on 80 acres of vegetables, Groff is considering a "soft," narrow-spectrum insecticide that targets slugs without threatening earthworms.

Weeds get the personal touch

Each of Groff's fields has its own "recipe" for weed control. On his four-wheeler, he diligently scouts his crops, searching for small weeds and weighing his options. "It's intensive management of weeds, and it's not a second or third priority — it's a top priority," he says.

To control weeds, Groff depends primarily on crop rotation and cover crops but he says the third component of his system — no-till — curtails their numbers to begin with. "The long-term effect of no-till is that you're not tilling up weed seeds, so if you keep up with the weeds, you can get away with not using as many herbicides." Although annual weeds aren't a problem on his farm, Groff says he frequently spot-treats perennial weeds. He has grown some crops without herbicides, but only when his cover crops smothered all of the weeds.

Adapting, innovating and learning for success

Because no-till soils are slower to heat up in the spring, Groff cleans off narrow bands where he will plant his sweet corn seed. By minimally tilling an area 6 to 8 inches wide and 3 inches deep, he fluffs up, dries and warms the soil right where the seed will be placed. By July, Groff's cooler no-till soils retain more moisture than tilled fields — an important asset in a region where summer drought is common. "In the beginning of the year, cooler soils are your enemy, but in the middle of the year they're your friend."

Groff protects his early tomatoes with high tunnels. To warm his sweet corn for 30 to 40 days in spring, he lays a clear, degradable plastic — developed in Ireland to extend dairies' field-corn season — over his rows immediately after planting. "We're able to get corn as early as anyone else in the area," he says, "but because it's clear plastic, it actually enhances weed growth, so I have to use normal herbicide rates there."

Steve Groff uses many little hammers, or strategies, to battle pests on his Pennsylvania farm.

Other innovations abound at Cedar Meadow Farm. Unable to find what he needed in the marketplace, Groff designed a no-till vegetable transplanter, uses a Buffalo rolling stalk chopper and modifies much of his other equipment.

His cover-crop mixes of hairy vetch,
crimson clover and rye … seem to obstruct, exhaust,
confuse and otherwise inhibit Colorado potato beetles.

While his system clearly presents challenges, its benefits overwhelm them: Groff says his farm's organic matter has increased from 2.7 percent to 4.8 percent, his soil microbial biomass has tripled and his soil aggregate stability has quadrupled. Over the years, his crop yields have improved — on average — about 10 percent.

Groff advises interested farmers to start out small and learn all they can. "There's a lot of art and technique to this way of farming," he says. "It may work right off the bat but it may take you a couple of years to learn how to use it. One of the biggest challenges is knowing how and why the system works."

The cover crop mixture of balansa clover, crimson clover (shown) and hairy vetch helped build beneficial insect populations early in the season.

As many as 120 species of beneficial arthropods have been found in southern Georgia soils when cotton residues were left on the surface and insecticides were not applied. In just one vegetable-growing season, 13 known beneficial insects were associated with cover crops. When eggplant was transplanted into crimson clover at 9 a.m., assassin bugs destroyed Colorado potato beetles on the eggplant by evening. Similarly, other beneficials killed cucumber beetles on cucumber plants within a day.

Underlying those benefits, according to the Georgia researchers, was the soil-improving combination of cover crops with conservation tillage:

TIP To support beneficial soil organisms, plant cover crops and allow their residue to accumulate on the soil surface.

soil organic matter increased from less than 1 percent to 3 to 8 percent in most of their plots, and a majority of growers saw similar improvements in soils and pest management.

Impacts of Fertilizers on Insect Pests

By modifying the nutrient composition of crops, fertilizer practices can influence plant defenses. A review of 50 years of research identified 135 studies showing more plant damage and/or greater numbers of leaf-chewing insects or mites in nitrogen-fertilized crops, while fewer than 50 studies reported less pest damage. Researchers have demonstrated that high nitrogen levels in plant tissue can decrease resistance and increase susceptibility to pest attacks (Table 2). Although more research is needed to clarify the relationships between crop nutrition and pests, most studies assessing the response of aphids and mites to nitrogen fertilizer have documented dramatic expansion in pest numbers with increases in fertilizer rates.

TABLE 2
Pest Populations Increase with Excess Nitrogen Fertility

COMMON NAME	GENUS AND SPECIES	CROP
European red mite	*Panonychus ulmi*	Apples
Two-spotted spider mite*	*Tetranychus telarius*	Apples, beans, peaches, tomatoes
Clover mite	*Bryobia praetiosa*	Beans, peaches
Greenhouse thrip	*Heliothrips haemorrhoidalis*	Beans
Green peach aphid*	*Myzus persicae*	Brussels sprouts, tobacco
Greenbug	*Schizaphis graminum*	Oats, rye
Corn leaf aphid	*Rhopalosiphum maidis*	Sorghum
Spotted alfalfa aphid	*Therioaphis maculate*	Alfalfa

* Photo p. 54.

Crops could be expected, therefore, to be less prone to insect pests and diseases where organic soil amendments are used, since these amendments usually result in lower concentrations of soluble nitrogen in plant tissue. Indeed, most studies documenting fewer insect pests in organic systems have attributed these reductions in part to lower nitrogen content in the crop tissues:

- In Japan, the density of whitebacked planthopper (*Sogatella furcifera*) immigrants in organic rice fields was significantly less than their density in conventional rice fields. Fewer adult females settled in the organic fields and fewer immatures survived, leading to smaller ensuing generations. These results have been partly attributed to lower nitrogen content in the organically farmed crops.
- In England, conventional winter wheat fields were plagued with more rose-grain aphids than their organic counterpart. Top-dressed in April with nitrogen, the plants treated with soluble synthetic fertilizers con-

tained higher levels of free protein amino acids in their leaves in June and attracted larger populations of aphids. Researchers concluded that the aphids found the conventionally grown wheat to be more palatable than the organically grown wheat.

- In Ohio greenhouse experiments, European corn borer females laid significantly more eggs on sweet corn growing in conventionally fertilized soils than they did on plants growing in organically farmed soils collected nearby. Interestingly, egg-laying varied significantly among the chemically fertilized treatments but was uniformly low in organically managed soils. The difference appears to be evidence for a form of biological buffering more commonly found under organic conditions.
- In California, organically fertilized broccoli consistently developed smaller infestations of flea beetles and cabbage aphids than conventionally fertilized broccoli. Researchers attributed those reduced infestations to lower levels of free nitrogen in plant foliage, further supporting the view that farmers can influence insect pest preferences with the types and amounts of fertilizers they use.
- In tropical Asia, by increasing organic matter in irrigated rice, researchers enhanced populations of decomposers and plankton-feeders — key components in the food chain of predators; in turn, numbers of generalist predators of leafhopper pests rose significantly. Organic matter management proved to underlie higher levels of natural biological control.

Implications for Fertilizer Practices

Conventional synthetic fertilizers can dramatically affect the balance of nutritional elements in plants. When farmers use them excessively, these fertilizers likely create nutritional imbalances with their large pulses of available nitrogen, which in turn compromise crops' resistance to insect pests.

In contrast, most organic farming practices lead to increased organic matter and microbial activity in soils and the gradual release of plant nutrients; in theory, this should provide more balanced nutrition to plants. While the amount of nitrogen that is immediately available to the crop may be lower when farmers use organic inputs, their crops' overall nutritional status appears to improve. By releasing nitrogen slowly, over the course of several years, organic sources may help render plants less attractive to pests. Organic soil fertility practices also can supply secondary and trace elements, such as boron, zinc, manganese and sulfur, which are occasionally lacking in conventional farming systems that rely primarily on synthetic sources of nitrogen, phosphorus and potassium.

If, indeed, biochemical or mineral-nutrient differences in organically grown crops enhance resistance, this may explain — at least in part — why lower pest levels have been reported in organic farming systems. Observations of these lower levels support the view that long-term management of soil organic matter leads to better plant resistance to insect pests.

At the USDA Beltsville Agricultural Research Center, researchers discovered a molecular basis for delayed leaf senescence and tolerance to diseases in tomato plants grown in a hairy vetch mulch, compared to the same crop grown on black plastic. The finding is an important step toward a scientific rationale for alternative soil management practices.

Probably due to regulated release of carbon and nitrogen metabolites from hairy vetch decomposition, the cover-cropped tomato plants showed a distinct expression of selected genes, which would lead to a more efficient utilization and mobilization of C and N, promote defense against disease, and enhance crop longevity. These results confirm that in intensive conventional tomato production, the use of legume cover crops offers advantages as a biological alternative to commercial fertilizer, in addition to minimizing soil erosion and loss of nutrients, enhancing water infiltration, reducing runoff, and creating a "natural" pest-predator relationship.

Traditionally considered in isolation from one another, aboveground and belowground components of ecosystems are now thought to be closely linked. The (crop) plant seems to function as an integrator of the above ground and below ground components of agroecosystems. This holistic approach is enhancing our understanding of the role of biodiversity at a global level. In agriculture, such close ecological linkages between aboveground and belowground biota constitute a key concept on which a truly innovative ecologically based pest management strategy can be built.

5 Beneficial Agents On the Farm

 BIOLOGICAL CONTROL is the use of natural enemies to manage pests. The natural enemy might be a predator, parasite, or disease that will attack the insect pest. Biological control is a form of enhancing natural defenses to achieve a desired effect. It usually involves raising and releasing one insect to have it attack another, almost like a "living insecticide." You can facilitate a biological control program by timing pesticide applications or choosing pesticides that will be least harmful to beneficial insects.

A durable biological control program depends on two main strategies:

1) Using ecological farm design to make your farm more attractive to biological control "agents."
2) Introducing beneficial agents onto your farm.

When plant pathogens are not inhibited by naturally occurring enemies, you can improve biocontrol by adding more effective beneficials. Such "directed biocontrol" operates in several ways. As naturally occurring enemies would do, introduced beneficials may:

- produce antibiotics
- parasitize target organisms
- form physical or chemical barriers to infection
- outcompete plant pathogens for niches
- simply help the plant grow better, masking symptoms where disease is present.

Predators

Predators occur in most orders of insects but primarily in the beetle, dragonfly, lacewing, wasp and true bug families (*Coleoptera, Odonata, Neuroptera, Hemiptera and Diptera,* respectively). Their impacts have been highlighted worldwide by eruptions of spider mite pests where chemical insecticides have eliminated the mites' predators. Tetranychid mites, for example, are usually very abundant in apple orchards where pesticides have destroyed natural predator populations.

The diversity of predator species in agroecosystems can be impressive. Researchers have reported more than 600 species — from 45 families — of predaceous arthropods in Arkansas cotton fields and about 1,000 species in Florida soybean fields. Such diversity can apply major regulatory pressures on pests. Indeed, many entomologists consider native, or indigenous, predators a sort of balance wheel in the "pest-natural enemy complex"

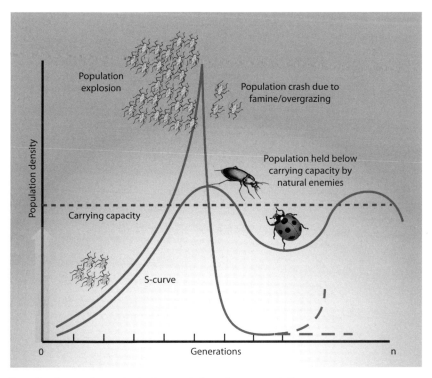

The role of natural enemies in the regulation of pest populations.

(Reprinted with permission from *Michigan Field Crop Pest Ecology,* Mich. State Univ. Extension Bulletin E-2704.)

TABLE 3
Common Predators*

COMMON NAME	ORDER	FAMILY	HOST OR PREY
Praying mantids	Orthoptera	Mantidae	Large and small insects
Earwigs	Dermaptera	Labiduridae	Caterpillars, many others
Predaceous thrips	Thysanoptera	Aeolothripidae	Spider mite eggs
Minute pirate bugs	Hemiptera	Anthocoridae	Insect eggs, soft-bodied insects, small insects
Big-eyed bugs		Lygaeidae	Insect eggs, soft-bodied insects, small insects
Plant bugs		Miridae	Insect eggs, soft-bodied insects, small insects
Damsel bugs		Nabidae	Insect eggs, small insects
Assassin bugs		Reduviidae	Small insects, caterpillars
Predaceous stink bugs		Pentatomidae	Small caterpillars
Lacewings	Neuroptera	Chrysopidae	Aphids, soft-bodied insects
Lady beetles	Coleoptera	Coccinellidae	Aphids, soft-bodied insects, insect eggs
Ground beetles		Carabidae	Insect eggs, soft-bodied insects, caterpillars
Rove beetles		Staphylinidae	Small insects
Soft-winged flower beetles		Melyridae	Insect eggs, soft-bodied insects, small caterpillars
Predaceous midges	Diptera	Cecidomyiidae	Aphids
Syrphid/hover flies		Syrphidae	Aphids, soft-bodied insects
Ants	Hymenoptera	Formicidae	Insect eggs, soft-bodied insects, small insects
Hornets, yellow jackets		Vespidae	Caterpillars, small insects
Digger wasps, mud daubers		Sphecidae	Caterpillars, small insects

* See Key to Beneficials and Pests, pp. 50–54.

because they tend to feed on whatever pest is over-abundant. Even where predators can't force pest populations below economically damaging levels, they can and do slow down the rate at which potential pests increase. In spray-free apple orchards in Canada, five species of predaceous true bugs were responsible for 44 to 68 percent of the mortality of codling moth eggs.

Biodiverse farms are rich in predatory insects, spiders and mites. These beneficial arthropods prey on other insects and spider mites and are critical to natural biological control (Table 3). They can feed on any or all stages of their prey, destroying or disabling eggs, larvae, nymphs, pupae or adults. Some predators — like lady beetles and ground beetles — use chewing mouthparts to grind up and bolt down their prey. Others — like assassin bugs, lacewing larvae and syrphid fly larvae — have piercing mouthparts; they often inject powerful toxins into their prey, quickly immobilizing them before sucking their juices.

Many predatory arthropods — including lady beetles, lacewing larvae and mites — are agile and ferocious hunters. They actively stalk their prey on the ground or in vegetation. Other hunters — such as dragonflies and robber flies — catch their prey in flight. In contrast, ambushers patiently sit and wait for mobile prey; praying mantids, for example, are usually well camouflaged and use the element of surprise to nab their unsuspecting victims.

Most predators are "generalist" feeders, attacking a wide variety of insect species and life stages. They may have preferences — lady beetles and lacewings, for instance, favor aphids — but most will attack many other prey that are smaller than themselves. Some important predator species are cannibalistic; green lacewings and praying mantids are notorious for preying on younger and weaker members of their own species. The diet of most predators also includes other beneficial insects, with larger predators frequently making meals of smaller predators and parasites.

As a rule, predators are predaceous regardless of their age and gender and consume pollen, nectar and other food as well as their prey. However, some species are predaceous only as larvae; as adults, they feed innocently on nectar and honeydew or aid and abet the predatory behavior of their young by laying their eggs among the prey. Lacewings are predaceous only during their immature stages. Other species are lifelong predators but change targets as they mature.

Principal Insect Predators

Spiders. Spiders are among the most neglected and least understood of predators. They rely on a complex diet of prey and can have a strong stabilizing influence on them. Because spiders are generalists and tend to kill more prey than they actually consume, they limit their preys' initial bursts of growth.

Many spiders live in crop canopies but most inhabit the soil surface and climb plants. Fields with either living plants or residue as soil cover tend to harbor diverse and abundant spider populations. Up to 23 spider families have been documented in cotton and 18 species have been tallied in apples. Because such diverse populations of spiders remain relatively constant, they maintain tolerable levels of their associated prey without extinguishing them.

TIP Fields with either living plants or residue as soil cover tend to harbor diverse and abundant spider populations. Living mulches composed of clover or other soil plant covers attract spiders, while residue from plants like barley or rye also harbor spider populations.

Lady beetles (Coccinellidae, also called ladybugs or ladybird beetles). With their shiny, half-dome bodies and active searching behavior, lady beetles are among the most visible and best known beneficial insects. More than 450 native or introduced species have been found in North America. They are easily recognized by their red or orange color with black markings, although some are black with red markings and others have no markings at all.

Lady beetles have been used in biological control programs for more than a century and are beneficial both as adults and larvae. Most larvae are blueblack and orange and shaped like little alligators. Young larvae pierce their prey and suck out their contents. Older larvae and adults chew entire aphids.

Any crop prone to aphid infestation will benefit from lady beetles, even though this predator's vision is so poor that it almost has to touch an aphid to detect it. Growers of vegetables, grains, legumes, strawberries and orchard crops have all found lady beetles helpful in managing aphids. In its lifetime, a single beetle can eat more than 5,000 aphids. In the Great Plains, studies of greenbug pests in grain sorghum have shown that each lady beetle adult can consume almost one of these aphids per minute and dislodge three to five times that many from the plant, exposing the dislodged greenbugs to ground-dwelling predators.

While their primary diet is aphids, lady beetles can make do with pollen, nectar and many other types of prey, including young ladybugs. Indeed, their extensive prey range — which includes moth eggs, beetle eggs, mites, thrips and other small insects — makes lady beetles particularly valuable as natural enemies.

Ground beetles (Carabidae). Predaceous ground beetles, or carabids, belong to a large family of beneficial beetles called the Carabidae whose adults live as long as two to four years. Several thousand species reside in North America alone.

Generally nocturnal, most predaceous ground beetles hide under plant litter, in soil crevices or under logs or rocks during the day. At night, their long, prominent legs allow them to sprint across the ground in pursuit of prey. Some species even climb up trees, shrubs or crops.

Most adult ground beetles range in length from 0.1 to 1.3 inches (3.2–32 mm). Their antennae are fairly threadlike and their bodies — although quite variable — are often heavy, somewhat flattened and either slightly or distinctly tapered at the head end. Some species are a brilliant or metallic purple, blue or green, but most are dark brown to black.

Armed with large, sharp jaws, adult predaceous ground beetles are ferocious. They can consume their body weight in food each day. Some carabids grind and eat such annual weed seeds as foxtail and velvetleaf. Larval carabids are not always predatory. In the *Lebia* genus, for example, adults are predators but first-instar larvae are parasites of chrysomelid beetles. (Instars are stages between successive molts.) Normally colorful, *Lebia* adults are just 0.1 to 0.6 inches (2.5–14 mm) long. *Lebia grandis* is a native and specialist predator of all immature stages of the Colorado potato beetle in cultivated potatoes in the eastern and mideastern U.S.

Lacewings (Chrysoperla spp.). Green lacewings — with their slender, pale-green bodies, large gauze-like wings and long antennae — are very common in aphid-infested crops, including cotton, sweet corn, potatoes, tomatoes, peppers, eggplants, asparagus, leafy greens, apples, strawberries and cole crops.

The delicate, fluttering adults feed only on nectar, pollen and aphid honeydew. About 0.5 to 0.8 inches (12–20 mm) long, they are active fliers — particularly during the evening and night, when their jewel-like golden eyes often reveal their presence around lights.

The larvae — tiny gray or brown "alligators" whose mouthparts resemble ice tongs — are active predators and can be cannibalistic. Indeed, green lacewing females suspend their oval eggs singly at the ends of long silken

COVER CROPS LURE BENEFICIAL INSECTS, IMPROVE BOTTOM LINE IN COTTON

SARE-funded researchers in Georgia seeking new ways to raise healthy cotton — traditionally one of the most pest-plagued, thus one of the most chemically treated commodities — focused on attracting insects that prey on damaging pests. A group of scientists from USDA's Agricultural Research Service, the University of Georgia, and Fort Valley State University planted a variety of flowering cover crops amid cotton rows to test whether their blooms would bring earworm- and budworm-killing predators to minimize the need for insecticides.

Working on seven mid- and southern Georgia cotton farms, the team eliminated one insecticide application by planting legume cover crop mixes that brought predators like the pirate bug, big-eyed bug and fire ants to prey on damaging worms. Using conservation tillage to plant cotton amid the cover crops also improved yields — on average, 2,300 pounds of seed cotton compared to 1,700 pounds on control plots. (Seed cotton weight includes lint and seed before cleaning.)

Growing a mix of balansa clover, crimson clover, and hairy vetch prolonged cover crop flowering from early March through late April and had the added benefit of out-competing weeds. "With this range of blooming, we're able to start building the beneficial populations early in the season," said Harry Schomberg, an ARS ecologist and project leader. "Reducing one application of insecticides could be pretty substantial on a larger scale like 100 acres." Glynn Tillman, an ARS entomologist who collaborated on the project, found that predator bugs moved from the cover crops into the cotton early in the season, providing more worm control. Moreover, the conservation tillage and cover crop residue resulted in more beneficial soil organisms that likely contributed to better cotton yields.

To demonstrate their results, the team went beyond holding field days. Tillman introduced the promising cotton-cover crop-conservation tillage system to hundreds of thousands attending the Sun Belt Ag Expo in Moultrie, Ga. "It was

stalks to protect them from hatching siblings. Commonly called aphid lions, lacewing larvae have well-developed legs with which to lunge at their prey and long, sickle-shaped jaws they use to puncture them and inject a paralyzing venom. They grow from less than 0.04 inch to between 0.2 and 0.3 inches (from <1 mm to 6–8 mm), thriving on several species of aphids as well as on thrips, whiteflies and spider mites — especially red mites. They will journey up to 100 feet in search of food and can destroy as many

well received," Tillman said, adding that she fielded many questions from growers, some calling later for more information on adopting cover crops into integrated pest management systems for cotton.

Schomberg cautions that the system requires careful management. In the fall, they seeded alternating strips of cereal rye and legume cover crops. In the spring, they killed the 15-inch-wide strips of rye with an herbicide and followed by planting cotton in the same rows, using conservation tillage. "Spacing is key," he said. "You have to think about and tinker with your planting equipment." Killing cover crops, he added, is easier than killing a diverse population of winter weeds.

Georgia researchers planted cotton into rows of legume cover crop mixes to attract insect predators to prey on damaging worms.

as 200 aphids or other prey per week. They also suck down the eggs of leafhoppers, moths and leafminers and reportedly attack small caterpillars, beetle larvae and the tobacco budworm.

Minute pirate bugs (Orius spp.). These often-underestimated "true bugs" are very small — a little over 0.1 inch (3 mm) long. The adults' white-patched wings extend beyond the tips of their black, somewhat oval bodies. The briskly moving nymphs are wingless, teardrop-shaped and yellow-orange to brown.

Minute pirate bugs are common on pasture, in orchards and on many agricultural crops, including cotton, peanuts, alfalfa, strawberries, peas, corn and potatoes. They feed greedily on thrips, insect eggs, aphids and small caterpillars and can devour 30 or more spider mites a day. Clasping their assorted small prey with their front legs, they repeatedly insert their needle-like beaks until they have drained their victims dry. They are prodigious consumers of corn earworm eggs in corn silks and also attack corn leaf aphids, potato aphids, potato leafhopper nymphs and European corn borers. Minute pirate bugs can even deliver harmless but temporarily irritating bites to humans.

Because they depend on pollen and plant juices to tide them over when their preferred prey are scarce, minute pirate bugs are most prevalent near spring- and summer-flowering shrubs and weeds.

Big-eyed bugs (Geocoris spp.). Named for their characteristically large, bulging eyes, big-eyed bugs are key and frequent predators in cotton and many other U.S. crops, including warm-season vegetables. *Geocoris punctipes* and *G. pallens* are the most common of the roughly 19 *Geocoris* species found in North America.

Adult big-eyed bugs — normally yellow or brown but sometimes black — are oval and small (0.12 to 0.16 inch, or 3–4 mm, long). Their unusually broad heads are equipped with piercing, sucking mouthparts. The similarly armed nymphs look like smaller, grayer versions of the adults.

Big-eyed bugs are omnivorous. Their diet includes plants but they prefer to prey on smaller insect and mite pests. They have been observed charging their intended victims, stabbing them quickly with their extended beaks and sometimes lifting them off the ground in the process.

Big-eyed bugs attack the eggs and small larvae of bollworm, pink bollworm and tobacco budworm and most other lepidopteran pests. They also target all life stages of whiteflies, mites and aphids and the eggs and nymphs of plant bugs. Laboratory studies indicate that a ravenous, growing nymph can exterminate 1,600 spider mites or about 250 soybean looper eggs before reaching maturity; adults have bolted down 80 spider mites or four lygus bug eggs a day.

Syrphid flies. Also known as hover flies because they hover and dart in flight, these brightly colored bee and wasp mimics are unusually voracious predators, as larvae, of aphids and other slow-moving, soft-bodied insects.

Depending on the species, many syrphid flies over-winter, giving rise to adults in spring. Adult syrphid flies feed on pollen, nectar and aphid honeydew. Each female lays hundreds of white, football-shaped eggs, about 0.04 inch (1 mm) long, amidst aphid colonies. The narrow, tapered slug-like larvae that hatch from these eggs can pierce and drain up to 400 aphids apiece during the two to three weeks it takes them to complete development. Unable to perceive their prey except through direct contact, syrphid fly larvae find their dinners by flinging their forward ends from side to side.

Parasitoids

Most parasitoids — parasitic insects that kill their hosts — live freely and independently as adults; they are lethal and dependent only in their immature stages. Parasitoids can be specialists, targeting either a single host species or several related species, or they can be generalists, attacking many types of hosts. Typically, they attack hosts larger than themselves, eating most or all of their hosts' bodies before pupating inside or outside them.

When the parasitoid emerges from its pupa as an adult, it usually nourishes itself on honeydew, nectar or pollen — although some adults make meals of their host's body fluids and others require additional water. Adult female parasitoids quickly seek out more victims in which to lay their host-killing eggs. With their uncanny ability to locate even sparsely populated hosts using chemical cues, parasitoid adults are much more efficient than predators at ferreting out their quarry.

Different parasitoids can victimize different life stages of the same host, although specific parasitoids usually limit themselves to one stage. Thus, parasitoids are classified as egg parasitoids, larval (nymphal) parasitoids or adult parasitoids. Some parasitoids lay their eggs in one life stage of a

TABLE 4
Common Parasitoids*

ORDER	FAMILY	HOST OR PREY	MODE OF ATTACK
Diptera (true flies)	Tachinidae	Beetles, butterflies, moths	Internal
	Nemestrinidae	Locusts, beetles	Internal
	Phoridae	Ants, caterpillars, termites, flies, others	Internal
	Crytochaetidae	Scale insects	Internal
Hymenoptera (ants, bees and wasps)	Chalcididae	Flies and butterflies (larvae and pupae)	Internal or external
	Encyrtidae	Aphids, scales, mealybugs, whiteflies	Internal
	Eulophidae	Aphids, gall midges, sawflies, mealybugs	Internal or external
	Pteromalidae	Flies, including houseflies and stable flies	Internal
	Pteromalidae	Boll weevils	External
	Aphelinidae	Whiteflies, scales, mealybugs, aphids	Internal or external
	Trichogrammatidae	Moth eggs	Internal
	Mymaridae	True bugs, flies, beetles, leafhopper eggs	Internal
	Scelionidae	Eggs of true bugs and moths	Internal
	Ichneumonidae	Larvae or pupae of beetles, caterpillars, wasps	Internal or external
	Braconidae	Larvae of beetles, caterpillars, sawflies	Internal (mostly)

* See Key to Beneficials and Pests, pp. 50–54.

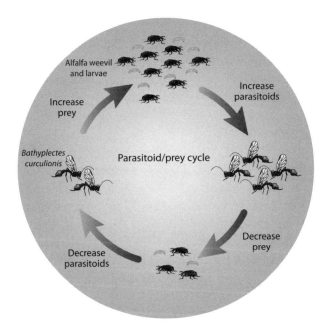

Populations of the ichneumonid parasitoid, *Bathyplectes curculionis,* and its prey, the alfalfa weevil, fluctuate throughout the growing season.

(Reprinted with permission from *Michigan Field Crop Pest Ecology,* Mich. State Univ. Extension Bulletin E-2704.)

victim but emerge at a later life stage. Parasitoids are also classified as either ectoparasites or endoparasites depending, respectively, on whether they feed externally on their hosts or develop inside them. Their life cycle is commonly short, ranging from 10 days to four weeks.

Principal Parasitoids

Dipteran flies. Entomologists have described more than 18,000 species of dipteran, or fly, parasites, which have diversified over an expansive range of hosts (Table 5). Unlike parasitic wasps, most species of parasitic flies lack a hardened structure with which to deposit eggs inside their hosts. Instead, they lay their eggs or larvae on plants, where the parasitoid larvae can easily penetrate the host but also where their target victims can eat them.

Individual species of parasitic flies are extraordinarily capable of surviving on many kinds of foods. The tachinid *Compsilura concinnata,* for example, has been successfully reared from more than 100 host species and three different host orders. Members of other Diptera families — such

TABLE 5
Major groups of dipteran (fly) parasitoids

FAMILY	DESCRIBED SPECIES (#)	PRIMARY HOSTS OR PREY
Sciomyzidae	100	Gastropods (snails/slugs)
Nemestrinidae	300	Orthoptera: Acrididae
Bombyliidae	5000	Primarily Hymenoptera, Coleoptera, Diptera
Pipunculidae	1000	Homoptera: Auchenorrhyncha
Conopidae	800	Hymenoptera: Adults of bumblebees and wasps
Sarcophagidae	750	Lepidoptera, Orthoptera, Homoptera, Coleoptera, Gastropoda + others
Tachinidae*	8200+	Lepidoptera, Hymenoptera, Coleoptera, Hemiptera, Diptera + many others
Pyrgotidae	350	Coleoptera: Scarabaeidae
Acroceridae	500	Arachnida: Araneae
Phoridae	300	Hymenoptera, Diptera, Coleoptera, Lepidoptera, Isoptera, Diplopoda + others
Rhinophoridae	90	Isopoda
Calliphoridae	240	Earthworms, gastropods

* Photo p. 53.

as big-headed flies in the *Pipunculidae* family, which are endoparasites of leafhoppers and planthoppers, and the small-headed *Acroceridae*, which only target spiders — are generally more specialized. However, some attack hosts from several families or subfamilies.

Chalcid wasps. For both natural and applied biological control, the chalcid wasps of the superfamily Chalcidoidea are among North America's most important insect groups. About 20 families and 2,000-plus species have been found on the continent — among the smallest of insects.

Because they are so diminutive, chalcid wasps are often underestimated in their numbers and effectiveness. They can be seen tapping leaf surfaces with their antennae in search of their host's "scent," but their presence is most commonly revealed by the sickly or dead hosts they leave in their wake. They parasitize a great number of pests, and different species attack different stages of the same host.

The following six families have proven especially useful in managing pests.

Fairyflies (Mymaridae). At between 0.008 and 0.04 inches (0.2–1 mm) long, these smallest of the world's insects can fly through the eye of a needle. Viewed under the microscope, the back wings of fairyflies contain distinctive long hairs.

Fairyflies parasitize the eggs of other insects — commonly flies, beetles, booklice and leafhoppers. Many fairyfly species, especially those belonging to the genus *Anaphes,* play crucial roles in biological control. The introduced egg parasite *A. flavipes,* for example, is one of two parasites that have been established for cereal leaf beetle management in small grains. In pesticide-free California vineyards with ground vegetation, the tiny *Anagrus epos* wasp can make a big dent in grape leafhopper densities.

Trichogramma wasps (Trichogrammatidae). Trichogramma wasps are the most widely released natural enemies. The tiny female wasp — generally less than 0.04 inch (1 mm) long — lays an egg inside a recently laid host egg, which blackens as the larva develops.

The host range of many Trichogramma wasps spans numerous species and families of insects. Moths, butterflies, beetles, flies, wasps and true bugs are all frequent victims. Some Trichogramma wasps even use their wings in a rowing motion to reach aquatic hosts.

Among commercially available species in the U.S. are *T. minutum, T. platneri* and *T. pretiosum,* which are released into fields on cards loaded with the parasitized eggs of non-pest hosts. Some foreign species — including *T. ostriniae, T. nubilale* and *T. brassicae* — also are being evaluated for augmentation biocontrol against European corn borers.

Eulophid wasps (Eulophidae). Eulophid wasps number more than 600 in North America, making theirs one of the largest chalcid families. About 0.04 to 0.12 inches (1–3 mm) long, they are often brilliant metallic blue or green.

Some species of eulophids are mite predators while others attack spider egg cases, scale insects and thrips. Most eulophids, however, parasitize flies, other wasps or the larvae or pupae of beetles or moths. Leaf-mining and wood-boring insects are frequent hosts.

Eulophids destroy many major crop pests. In the Midwest alone, *Sympiesis marylandensis* is an important parasite of spotted tentiform leafminer in apples. *Diglyphus isaea* — available commercially — is a primary parasite of agromyzid leafminers in greenhouses. *Edovum puttleri* attacks the eggs of Colorado potato beetles. Finally, *Pediobius foveolatus* — introduced from India and also available commercially — parasitizes Mexican bean beetle larvae.

Pteromalid wasps (Pteromalidae). This large family of wasps assaults many types of insects, including the larvae of moths, flies, beetles and wasps. Several pteromalids target scale insects and mealybugs and some even act as "hyperparasitoids" — parasitizing other parasites within their hosts.

In the upper Midwest, *Pteromalus puparum* is a key enemy of imported cabbageworm pupae, each of which can involuntarily host more than 200 *Pteromalus* offspring. *Anisopteromalus calandrae*, which attacks the larvae of beetles that infest stored grain, impressed scientists several decades ago with its ability to suppress 96 percent of rice weevils in wheat spillage in small rooms. *A. calandrae* can now be purchased for release in grain storage and handling facilities.

Encyrtid wasps (Encyrtidae). Responsible for much of the classical biological control of scale insects and mealybugs in fruit trees, this important family of natural enemies encompasses about 400 species in the U.S. and Canada. Its extensive host range includes soft scales, armored scales, mealybugs and the eggs or larvae of insects in about 15 families of beetles, 10 families of flies and 20 families of moths and butterflies.

Several commercially available encyrtids now help manage scale and mealybugs in greenhouses: *Leptomastix dactylopii,* for example, parasitizes citrus mealybug, while *Metaphycus helvolus* attacks black, hemispherical, nigra, citricola, brown soft and other soft scales.

Other noteworthy encyrtids include *Ooencyrtus kuwanae*, an introduced parasite of gypsy moth eggs, and *Copidosoma floridanum*, a native parasite of cabbage looper larvae.

Aphelinid wasps (Aphelinidae). The effectiveness of aphelinids in managing scale insects has earned them one of the best reputations in biological control. They also destroy mealybugs, whiteflies, aphids and other families of Homoptera.

Aphelinus varipes parasitizes greenbugs, *A. mali* targets the woolly apple aphid, and members of the genus *Eretmocerus* attack silverleaf whitefly. *Encarsia formosa,* in commercial use since the 1920s, is now released into greenhouses worldwide; it kills almost 100 greenhouse whitefly nymphs during its 12-day life span.

Principal Insect Pathogens

Just like humans and other vertebrates, insects are susceptible to many disease-causing organisms known as pathogens. Thousands of species of bacteria, fungi, viruses, protozoa and nematodes can sicken or kill insects. Even if the insects survive, the pathogens' "sub-lethal" effects can keep their victims from feeding or reproducing.

Bacteria. Most bacteria infect specific insect orders. Some naturally occurring insect-pathogenic bacteria have been isolated and mass-produced for commercial use. One of these, *Bacillus thuringiensis* or Bt, is the world's most widely applied biological control agent. It exerts its toxicity only after plant-eating insects actually consume it. A highly dense protein crystal, the Bt toxin kills victims by first paralyzing their mid-gut, then their entire bodies. Like most other bacterial pathogens, Bt is specific to certain insect orders. Its short residual period also makes it an ideal candidate for pest management in fruits and vegetables.

Fungi. Although an estimated 700-plus species of fungi can infect insects, fewer than 20 have been developed for insect management. Most insect-pathogenic fungi need cool, moist environments to germinate. Compared to most other insect pathogens, they have an extensive host range. *Beauveria bassiana,* for example, can help manage beetles, ants, termites, true bugs, grasshoppers, mosquitoes and mites as well as other arthropod pests. It unleashes a toxin that weakens its host's immune system, then overwhelms its dead host's intestinal bacteria with an antibiotic. The tell-tale sign of *B. bassiana's* carnage is its victim's "white bloom" of fungal spores.

Fungi can invade their insect host through natural openings in its cuticle. Thus, hosts need not consume pathogens but only come into direct contact with them. Although some fungi can take up to several weeks to kill their hosts, most infected insects die within three to seven days.

Viruses. Most viruses that attack insects belong to a group called nuclear polyhedrosis viruses or NPVs. Their victims are usually young larvae of butterflies and moths, which become infected by eating NPV particles and typically die within several weeks. Some infected larvae hang limply from the tops of crop canopies, prompting the common name "caterpillar wilt" or "tree top" disease.

Prevailing environmental factors heavily influence the insect-killing efficiencies of viruses. For example, they are adversely affected by sunlight, while the relatively slow speed at which they kill has also hindered their widespread acceptance for biocontrol.

Nematodes. Nearly 40 known families of nematodes parasitize and consume insects and other arthropods. Some are hunter-cruisers while others are ambushers. The most beneficial of these "entomopathogenic" nematodes belong to the *Heterorhabditidae* and *Steinernematidae* families. Both families are "obligate" parasites: their survival depends on their hosts and on the symbiotic relationships the nematodes have evolved with disease-causing *Xenorhabdus* and *Photorhabdus* bacteria.

Parasitic nematodes transport bacteria inside their host, penetrating the host via the mouth, anus, spiracles or cuticle. Once inside, the nematodes release the bacteria, which quickly multiply and kill the host. In turn, the nematode uses the bacteria and insect cadaver for food and shelter, maturing, mating and reproducing inside it. Infective-stage juvenile nematodes eventually emerge from the cadaver and seek out another host.

Because they are highly mobile and can locate and destroy new victims in just a few days, entomopathogenic nematodes make outstanding candidates for all kinds of biological control. Some are applied to soils to successfully manage the underground life stages of insect pests.

6 Putting It All Together

AGROECOLOGY — the science that underlies sustainable farming — integrates the conservation of biodiversity with the production of food. It promotes diversity which in turn sustains a farm's soil fertility, productivity and crop protection.

Innovative approaches that make agriculture both more sustainable and more productive are flourishing around the world. While trade-offs between agricultural productivity and biodiversity seem stark, exciting opportunities for synergy arise when you adopt one or more of the following strategies:

- Modify your soil, water and vegetative resource management by limiting external inputs and emphasizing organic matter accumulation, nutrient recycling, conservation and diversity.
- Replace agrichemical applications with more resource-efficient methods of managing nutrients and pest populations.
- Mimic natural ecosystems by adopting cover crops, polycultures and agroforestry in diversified designs that include useful trees, shrubs and perennial grasses.
- Conserve such reserves of biodiversity as vegetationally rich hedgerows, forest patches and fallow fields.
- Develop habitat networks that connect farms with surrounding ecosystems, such as corridors that allow natural enemies and other beneficial biota to circulate into fields.

Growing rye between vineyard rows suppresses weeds and attracts beneficial insects such as lady beetles to this Monterey County, Calif., vineyard.

Different farming systems and agricultural settings call for different combinations of those key strategies. In intensive, larger-scale cropping systems, eliminating pesticides and providing habitat diversity around field borders and in corridors are likely to contribute most substantially to biodiversity. On smaller-scale farms, organic management — with crop rotations and diversified polyculture designs — may be more appropriate and effective. Generalizing is impossible: Every farm has its own particular features, and its own particular promise.

Designing a Habitat Management Strategy

The most successful examples of ecologically based pest management systems are those that have been derived and fine-tuned by farmers to fit their particular circumstances. To design an effective plan for successful habitat management, first gather as much information as you can. Make a list of the most economically damaging pests on your farm. For each pest, try to find out:

- What are its food and habitat requirements?
- What factors influence its abundance?
- When does it enter the field and from where?

- What attracts it to the crop?
- How does it develop in the crop and when does it become economically damaging?
- What are its most important predators, parasites and pathogens?
- What are the primary needs of those beneficial organisms?
- Where do these beneficials over-winter, when do they appear in the field, where do they come from, what attracts them to the crop, how do they develop in the crop and what keeps them in the field?
- When do the beneficials' critical resources — nectar, pollen, alternative hosts and prey — appear and how long are they available? Are alternate food sources accessible nearby and at the right times? Which native annuals and perennials can compensate for critical gaps in timing, especially when prey are scarce?
- See Resources p. 104 and/or contact your county extension agent to help answer these questions.

CAUTION! Converting to organic production is no guarantee that your fields will be pest-free, even if you surround them with natural vegetation. Pest levels are site-specific: they depend on which plants are present, which insects are associated with them and how you manage both.

The examples below illustrate specific management options to address specific pest problems:

- In England, a group of scientists learned that important beneficial predators of aphids in wheat over-wintered in grassy hedgerows along the edges of fields. However, these predators migrated into the crop too late in the spring to manage aphids located deep in the field. After the researchers planted a 3-foot strip of bunch grasses in the center of the field, populations of over-wintering predators soared and aphid damage was minimized.
- Many predators and parasites require alternative hosts during their life cycles. *Lydella thompsoni,* a tachinid fly that parasitizes European corn borer, emerges before corn borer larvae are available in the spring and completes its first generation on common stalk borer instead. Clean farming practices that eliminate stalk borers are thought to contribute to this tachinid fly's decline.

- Alternative prey also may be important in building up predator numbers before the predator's target prey — the crop pest — appears. Lady beetles and minute pirate bugs can eventually consume many European corn borer eggs, but they can't do it if alternative prey aren't available to them before the corn borers lay their eggs.

- High daytime soil temperatures may limit the activity of ground-dwelling predators, including spiders and ground beetles. Cover crops or intercrops may help reduce soil temperatures and extend the time those predators are active. Crop residues, mulches and grassy field borders can offer the same benefits. Similarly, many parasites need moderate temperatures and higher relative humidity and must escape fields in the heat of day to find shelter in shady areas. For example, a parasitic wasp that attacks European corn borers is most active at field edges near woody areas, which provide shade, cooler temperatures and nectar-bearing or honeydew-coated flowering plants.

Enhancing Biota and Improving Soil Health

Managing soil for improved health demands a long-term commitment to using combinations of soil-enhancing practices. The strategies listed below can aid you in inhibiting pests, stimulating natural enemies and — by alleviating plant stress — fortifying crops' abilities to resist or compete with pests.

- Add plentiful amounts of organic materials from cover crops and other crop residues as well as from off-field sources like animal manures and composts. Because different organic materials have different effects on a soil's biological, physical and chemical properties, be sure to use a variety of sources. For example, well-decomposed compost may suppress crop diseases, but it does not enhance soil aggregation in the short run. Dairy cow manure, on the other hand, rapidly stimulates soil aggregation.
- Keep soils covered with living vegetation and/or crop residue. Residue protects soils from moisture and temperature extremes. For example, residue allows earthworms to adjust gradually to decreasing temperatures, reducing their mortality. By enhancing rainfall infiltration, residue also provides more water for crops.
- Reduce tillage intensity. Excessive tillage destroys the food sources and micro-niches on which beneficial soil organisms depend. When you reduce your tillage and leave more residues on the soil surface, you create a more stable environment, slow the turnover of nutrients and encourage more diverse communities of decomposers.

Greg Porter, Univ. of Maine

Compost, judiciously applied, can replace mineral fertilizers and feed beneficial soil organisms.

Preston Roland

Frank Bibin of Quitman, Ga., checks a house he built in his pecan orchard to attract predatory wasps.

- Adopt other practices that reduce erosion, such as strip cropping along contours. Erosion damages soil health by removing topsoil that is rich in organic matter.
- Alleviate the severity of compaction. Staying off soils that are too wet, distributing loads more uniformly and using controlled traffic lanes — including raised beds — all help reduce compaction.
- Use best management practices to supply nutrients to plants without polluting water. Make routine use of soil and plant tissue tests to determine the need for nutrient applications. Avoid applying large doses of available nutrients — especially nitrogen — before planting. To the greatest extent possible, rely on soil organic matter and organic amendments to supply nitrogen. If you must use synthetic nitrogen fertilizer, add it in smaller quantities several times during the season. Once soil tests are in the optimal range, try to balance the amount of nutrients supplied with the amount used by the crops.
- Leave areas of the farm untouched as habitat for plant and animal diversity.

Individual soil-improving practices have multiple effects on the agroecosystem. When you use cover crops intensively, you supply nitrogen to the following crop, soak up leftover soil nitrates, increase soil organisms and improve crop health. You reduce runoff, erosion, soil compaction and plant-parasitic nematodes. You also suppress weeds, deter diseases and inoculate future crops with beneficial mycorrhizae. Flowering cover crops also harbor beneficial insects.

Strategies for Enhancing Plant Diversity

As described, increasing above-ground biodiversity will enhance the natural defenses of your farming system. Use as many of these tools as possible to design a diverse landscape:

- Diversify enterprises by including more species of crops and livestock.
- Use legume-based crop rotations and mixed pastures.
- Intercrop or strip-crop annual crops where feasible.
- Mix varieties of the same crop.
- Use varieties that carry many genes — rather than just one or two — for tolerating a particular insect or disease.
- Emphasize open-pollinated crops over hybrids for their adaptability to local environments and greater genetic diversity.
- Grow cover crops in orchards, vineyards and crop fields.
- Leave strips of wild vegetation at field edges.
- Provide corridors for wildlife and beneficial insects.
- Practice agroforestry, combining trees or shrubs with crops or livestock to improve habitat continuity for natural enemies.
- Plant microclimate-modifying trees and native plants as windbreaks or hedgerows.
- Provide a source of water for birds and insects.
- Leave areas of the farm untouched as habitat for plant and animal diversity.

As you work toward improved soil health and pest management, don't concentrate on any one strategy to the exclusion of others. Instead, combine as many strategies as make sense on your farm. Nationwide, producers are finding that the triple strategies of good crop rotations, reduced tillage and routine use of cover crops impart many benefits. Adding other strategies — such as animal manures and composts, improved nutrient management and compaction-minimizing techniques — provides even more.

*Nationwide, producers are finding that the triple strategies
of good crop rotations, reduced tillage and routine use of
cover crops impart many benefits.*

Rolling out your Strategy

Once you have a thorough knowledge of the characteristics and needs of
key pests and natural enemies, you're ready to begin designing a habitat-
management strategy specifically for your farm.

- Choose plants that offer multiple benefits — for example, ones that
  improve soil fertility, weed suppression and pest regulation — and
  that don't disrupt desirable farming practices.
- Avoid potential conflicts. In California, planting blackberries around
  vineyards boosts populations of grape leafhopper parasites but can
  also exacerbate populations of the blue-green sharpshooter that
  spreads the vinekilling Pierce's disease.
- In locating your selected plants and diversification designs over space
  and time, use the scale — field- or landscape-level — that is most
  consistent with your intended results.
- And, finally, keep it simple. Your plan should be easy and inexpensive
  to implement and maintain, and you should be able to modify it as
  your needs change or your results warrant.

In this book, we have presented ideas and principles for designing and
implementing healthy, pest-resilient farming systems. We have explained
why reincorporating complexity and diversity is the first step toward sustainable pest management. Finally, we have described the pillars of agroecosystem health (Figure 1, p. 9):

- Fostering crop habitats that support beneficial fauna
- Developing soils rich in organic matter and microbial activity

Throughout, we have emphasized the advantages of polycultures over
monocultures and, particularly, of reduced- or no-till perennial systems
over intensive annual cropping schemes.

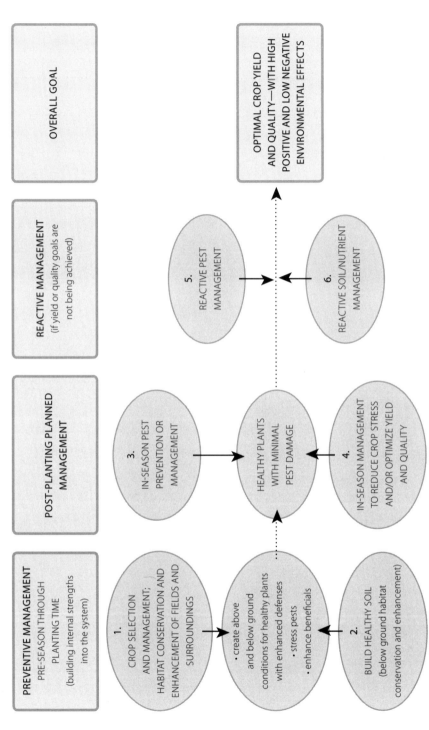

Figure 2. Preventive and reactive strategies that enhance ecological pest management. Adapted from Univ. of Vt., Dept. of Plant and Soil Sciences

Key Elements of Ecological Pest Management

Ecological Pest Management relies on *preventive* rather than *reactive* strategies. Your cropping program should focus primarily on preventive practices above and below ground (#1 and #2) to build your farm's natural defenses. Reactive management (#5 and #6) is reserved for problems not solved by the preventive or planned (#3 and #4) strategies.

OVERALL STRATEGIES:

— **Build the strengths of natural systems into your agricultural landscape to enhance its inherent pest-fighting capacity.**

— **Enhance the efficiency of your farm, including cycling of nutrients, flow of energy, and/or the use of other resources.**

These broad strategies and the individual practices that follow result in systems that are:

• Self-regulating — keeping populations of pests within acceptable boundaries
• Self-sufficient — with minimal need for "reactive" interventions
• Resistant to stresses such as drought, soil compaction, pest invasions
• Resilient -- with the ability to bounce back from stresses

1) Crop management: above ground habitat conservation and enhancement of biodiversity within and surrounding crop fields. Use a variety of practices or strategies to maintain biodiversity, stress pests and/or enhance beneficial organisms.

• Select appropriate crops for your climate and soil
• Choose pest resistant, local varieties and well adapted cultivars
• Use legume-based crop rotations, alternating botanically unrelated crops
• Use cover crops intensively
• Manage field boundaries and in-field habitats (ecological islands) to attract beneficials, and trap or confuse insect pests
• Use proper sanitation management
• Consider intercropping and agroforestry systems

2) Soil management: below ground habitat conservation and enhancement. Build healthy soil and maintain below ground biodiversity to stress pests, enhance beneficials and/or provide the best possible chemical, physical, and biological soil habitat for crops.

Build and maintain soil organic matter with crop residues, manures and composts
- Reduce soil disturbance (tillage)
- Keep soil covered with crop residue or living plants
- Use cover crops routinely
- Use longer crop rotations to enhance soil microbial populations and break disease, insect and weed cycles
- Maintain nutrient levels that are sufficient for crops but do not cause imbalances in the plant, which can increase susceptibility to insects and diseases
- Maintain appropriate pH
- Control soil erosion and nutrient losses
- Avoid practices that cause soil compaction

3) Planned supplemental pest management practices. The following practices can be used if research and farmer experience indicate that – despite the use of comprehensive preventive management as outlined above – some additional specific pest management practices will still be needed:

- Release beneficial insects or apply least environmentally harmful biopesticides
- Prune to reduce humidity in the canopy and deter fungal infections
- Cultivate for weed control based on knowledge of critical competition period

4) Planned supplemental soil practices to reduce crop stress and/or optimize yield and quality

- Maintain adequate soil water content (i.e., with careful irrigation scheduling)
- Mow rather than incorporate orchard cover crops, leaving a mulch cover
- Undersow legumes in cereals

5) Reactive inputs for pest management
If, after following preventive and planned management practices (#1, 2, 3, and 4), pests are above threshold levels and beneficials populations are low, release beneficials or apply selected biopesticides with low environmental impact.

6) Reactive inputs to reduce plant stress
- Use chisel plow or subsoiler to alleviate soil compaction
- Apply nutrients to soil or foliage in response to plant deficiency symptoms

FARM FEATURE

ROTATION, ROTATION, ROTATION:
ALFALFA, CLOVER CROPS BREAK PEST CYCLES

- Uses crop rotations to diversify soil biology and to thwart pests
- Provides habitat for beneficials
- Uses green manures to manage weeds

In Big Sandy, Mont., Bob Quinn hasn't borrowed operating capital from the bank for 10 years. Without hefty bills for agrichemicals at planting — and with an effective year-round marketing program — his cash flow is more stable than it was in the mid-1980s, before he began converting his 3,000 acres to organic.

The north central Montana dryland farm sells its organic barley, buckwheat and wheats — hard red winters, durums and hard red and soft white springs — for at least 50 percent more, on average, than conventional farms do. It also produces organic lentils and — under Quinn's Kamut® brand — the ancient Egyptian wheat khorasan. With fewer inputs and higher-value outputs, Quinn added a partner — Thad Willis — and another thousand acres. The expanded operation, now farmed entirely by Willis, supports two families instead of one. "That's a different direction than most of agriculture is going," says Quinn.

Indeed.

Rotation, rotation, rotation

Quinn attributes the farm's profitability to its soil-building, pest-thwarting, four- to five-year rotations. Its alfalfa, clover and grains are thick with predaceous lady beetles, and its last serious insect infestation was 15 years ago. "Most people can't believe it," he says. "For many years, people thought I was spraying at night. They couldn't believe anyone could succeed without chemicals."

Similarly, the viral diseases and root rots that used to sicken the farm's grains are "mostly gone," and pathogens flare only in the rare year when pre-harvest rains fuel black tip fungus in highly susceptible khorasan fields.

The farm's green-manure based weed-control program "works as well as conventional spraying," Quinn says: kochia has nearly disappeared ("I think it needs highly soluble nitrogen to compete with wheat"), wild oat seeds germinate more sparsely ("They're a problem for us, but not nearly as much as you would expect with no chemicals") and thistle is contained. Fanweed and mustards — which the partners unfortunately see more often now than they used to — succumb to the cultivator or to switchbacks between spring to fall planting.

"For many years, people thought I was spraying at night. They couldn't believe anyone could succeed without chemicals."

With no large livestock operations nearby to supply manure, the farm's "primary and only" instrument of soil improvement is green manure. In high-moisture years, that means weed-throttling alfalfa — underseeded in a grain crop the first year, hayed the second year and incorporated into the soil the third year. In intermediate-moisture years, the partners plant less-thirsty sweet clover with a companion grain the first growing season and disk or plow it under the second. In really dry years, they sow green-manure peas in the fall or green-manure lentils in the spring, turning them under by the first of June.

"I think the rotation and soil-building program we have in place allows a great diversity in soil biology, and that's what keeps the pests in place," Quinn says.

In their grain storages, Quinn and Willis dissuade pests with cool, drying air and with a dusting of insect-shredding diatomaceous earth laced with tempting pheromones. After they load their grain into bins, they level off the cones to eliminate peaks in which pest-supporting moisture and heat can concentrate.

Kamut®: World markets for local product

Because of his frequent and direct contacts with consumers, Quinn says he no longer thinks of himself as a commodity producer but instead as a grower of life- and health-sustaining food. He promotes and researches his Kamut® wheat worldwide when he's not developing a 100-megawatt wind farm in central Montana. Khorasan wheats appeal to consumers who are allergic to other wheats or who value low gly-

Bob Quinn's last serious insect infestation was 15 years ago.

cemic indices and high concentrations of antioxidants. They are used in more than 400 kinds of products — primarily cereals in the U.S., breads in northern Europe and tarts, pastries and pastas in southern Europe.

The Egyptian government has discouraged production of that nation's native khorasan wheat because of its low yield potential — a problem in Egypt's high-input, modern irrigation systems. That's not an issue in north central Montana, where dryland fields don't have high yield potential to begin with. That's precisely why khorasan is such a good fit, Quinn says. A 500-mile diameter area carved out of north central and northeastern Montana, southern Saskatchewan and southern Alberta is also the least likely to get rain when Kamut® wheat is most vulnerable to black tip — a dark discoloration of the germ end of otherwise healthy wheat kernels.

Rather than try to develop resistance to black tip, Quinn has used a post-harvest color sorter to "knock out the worst of it" when it occurs. "I'm not sure I want to breed in resistance," he says. "We like the wheat the way it is and don't want to take a chance on losing any of its wonderful qualities. So we've chosen not to tamper with it, to grow it in regions of the world where it's most successful and to be satisfied with lower yields." Besides, he notes, the lower numbers of bushels can be offset with higher prices.

Real results, real independence, real fun

Despite the partners' profitability, Quinn says only a handful of farmers in their area have adopted similar practices. "It's hard for a lot of people to change what they're doing," he says. "There are a lot of unknowns in this, and there's also a transition period when you will certainly experience lower yields without getting the organic premium." There's another reason, too: The word being spread by agrichemical companies

— and still coming down through traditional educational circles — is "that this it not real."

It's real all right, says Quinn, but it's certainly not real easy. "It takes a lot more management and thinking ahead. So if you aren't careful with your weeds, you can easily let them get ahead of you, and if you aren't careful with your rotations, the system won't work properly."

Farmers who make it work, however, find they are working directly for their customers rather than for Uncle Sam. "It puts us in a position to be paid a livelihood by the consumers rather than relying on government payments — and I think that's a very big plus," says Quinn.

Besides, he adds, "It really brought the fun back into farming."

Universal Principles, Farm-Specific Strategies

The key challenge for farmers in the 21st century is to translate the principles of agroecology into practical systems that meet the needs of their farming communities and ecosystems. You can apply these principles through various techniques and strategies, each of which will affect your farm differently, depending on local opportunities and resources and, of course, on markets. Some options may include both annual and perennial crops, while others do not. Some may transcend field and farm to encompass windbreaks, shelterbelts and living fences. Well-considered and well-implemented strategies for soil and habitat management lead to diverse and abundant — although not always sufficient — populations of natural enemies.

GUIDELINES FOR DESIGNING HEALTHY AND PEST-RESILIENT FARMING SYSTEMS

- Increase species in time and space with crop rotations, polycultures, agroforestry and crop-livestock systems.
- Expand genetic diversity with variety mixtures, local germplasm and multilines (or varieties that contain several different genes for resistance to a particular pest). In each case, the crop represents a genetically diverse array that can better withstand disease and pests.
- Conserve or introduce natural enemies and antagonists with habitat enhancement or augmentative releases.
- Boost soil biotic activity and improve soil structure with regular applications of organic matter.
- Enhance nutrient recycling with legumes and livestock.
- Maintain vegetative cover with reduced tillage, cover crops or mulches.
- Enhance landscape diversity with biological corridors, vegetationally diverse crop-field boundaries or mosaics of agroecosystems.

As you develop a healthier, more pest-resilient system for your farm, ask yourself:

- How can I increase species diversity to improve pest management, compensate for pest damage and make fuller use of resources?
- How can I extend the system's longevity by including woody plants that capture and recirculate nutrients and provide more sustained support for beneficials?
- How can I add more organic matter to activate soil biology, build soil nutrition and improve soil structure?
- Finally, how can I diversify my landscape with mosaics of agroecosystems in different stages of succession?

Because locally adapted varieties and species can create specific genetic resilience, rely on local biodiversity, synergies and dynamics as much as you can. Use the principles of agroecology to intensify your farm's efficiency, maintain its productivity, preserve its biodiversity and enhance its self-sustaining capacity.

10 Indicators of Soil Quality

Assign a value from 1 to 10 for each indicator, and then average all 10 indicators. Farms with overall values lower than 5 in either soil quality or crop health are considered below the threshold of sustainability and in need of rectifying measures.

INDICATOR	ESTABLISHED VALUES*	CHARACTERISTICS
Structure	1	Loose soil with no visible aggregates
	5	A few aggregates that break with little pressure
	10	Well-formed aggregates that break with difficulty
Compaction/ Infiltration	1	Compacted soil; accumulating water
	5	A thin compacted layer; slowly infiltrating water
	10	No compaction; easily infiltrating water
Soil depth	1	Exposed subsoil
	5	A thin layer of superficial soi
	10	Superficial soil that is >4 inches (10 cm.) deep
Status of residues	1	Slowly decomposing organic residues
	5	Last year's decomposing residues still present
	10	Residues in various stages of decomposition or all residues well-decomposed
Color, odor and organic matter	1	Pale; chemical odor; no humus
	5	Light brown; odorless; some humus
	10	Dark brown; fresh odor; abundant humus
Water retention (moisture level)	1	Dry soil
	5	Limited moisture
	10	Reasonable moisture
Root development	1	Poorly developed; short roots
	5	Roots with limited growth; some fine roots
	10	Healthy, well-developed roots; abundant fine roots
Soil cover	1	Bare soil
	5	<50% covered with residues or live cover
	10	>50% covered with residues or live cover
Erosion	1	Severe, with small gullies
	5	Evident but with few signs
	10	No major signs
Biological activity	1	No signs
	5	A few earthworms and arthropods
	10	Abundant organisms

*1=least desirable, 5=moderate, 10=preferred.

10 Indicators of Crop Health

INDICATOR	ESTABLISHED VALUES*	CHARACTERISTICS
Appearance	1	Chlorotic, discolored foliage with signs of deficiency
	5	Light-green foliage with some discoloring
	10	Dark-green foliage with no signs of deficiency
Crop growth	1	Poor growth, short branches, limited new growth, sparse stand
	5	Denser but not uniform stand, thin branches, some new growth
	10	Dense, uniform stand with vigorous growth
Tolerance or resistance to stress	1	Susceptible; does not recover well after stress
	5	Moderately susceptible; recovers slowly after stress
	10	Tolerant; recovers quickly after stress
Disease or pest incidence	1	Susceptible; >50% of plants damaged
	5	20–45% of plants damaged
	10	Resistant; <20% of plants with light damage
Weed competition and pressure	1	Crops stressed and overwhelmed by weeds
	5	Moderate presence of weeds exerting some competition
	10	Vigorous crop that overcomes weeds
Actual or potential yield	1	Low in relation to local average
	5	Medium or acceptable in relation to local average
	10	Good or high in relation to local average
Genetic diversity	1	One dominant variety
	5	Two varieties
	10	More than two varieties
Plant diversity	1	Monoculture
	5	Two species
	10	More than two species
Natural surrounding vegetation	1	Surrounded by other crops; no natural vegetation
	5	Adjacent to natural vegetation on at least one side
	10	Adjacent to natural vegetation on at least two sides
Management system	1	Conventional agrichemical inputs
	5	In transition to organic; IPM or input substitution
	10	Diversified; organic inputs; low external inputs

*1=least desirable, 5=moderate, 10=preferred.

Resources

General Information

Sustainable Agriculture Research and Education (SARE) program, USDA-CSREES, Washington, D.C. Studies and spreads information about sustainable agriculture via a nationwide grants program. See research findings at www.sare.org/projects

Sustainable Agriculture Network, Beltsville, Md. The national outreach arm of SARE, SAN disseminates information through electronic and print publications, including:

  –*Building Soils for Better Crops, 2nd Edition.* $19.95 + $5.95 s/h.
  – *Managing Cover Crops Profitably, 2nd Edition.* $19 + $5.95 s/.
  – *Steel in the Field: A farmer's guide to weed management tools.* $18 + $5.95 s/h.

To order: www.sare.org/WebStore; (301) 374-9696

Alternative Farming Systems Information Center (AFSIC), National Agricultural Library, Beltsville, Md. Offers bibliographic reference publications on ecological pest management online. (301) 504-6559; afsic@nal.usda.gov; www.nal.usda.gov/afsic

Appropriate Technology Transfer for Rural Areas (ATTRA), Fayetteville, Ark. Offers a series of publications on agronomy and pest management covering various aspects of ecological pest management. (800) 346-9140; http://attra.ncat.org

Publications

AgroEcology: Ecological Processes in Sustainable Agriculture by Stephen R. Gliessman. 1998. Sleeping Bear Press/Ann Arbor Press. http://www. agroecology.org/textbook.htm

Agroecology: The Science of Sustainable Agriculture (2nd ed.) by Miguel Altieri. Key principles in case studies of sustainable rural development in developing countries. $39 to Perseus Books Group Customer Service, (800) 371-1669; perseus.orders@perseusbooks.com; www.westview-press.com

Agroecology: Transitioning organic agriculture beyond input substitution by Miguel A. Altieri and Clara I. Nicholls. Paper presented to the American Society of Agronomy, November, 2003, Denver, CO. http://www. misa.umn.edu/Other/symposium/Altieri%20Paper.pdf

An alternative agriculture system is defined by a distinct expression profile of select gene transcripts and proteins. 2004. V Kumar, DJ Mills, JD Anderson, and AK Mattoo PNAS 101(29): 10535-10540. http://www.pnas. org/cgi/content/full/0403496101

Alternatives in Insect Pest Management — Biological and Biorational Approaches by University of Illinois Extension. Rates the effectiveness of microbial insecticides, botanical insecticides soaps, attractants, traps, beneficial insects, etc. Web only. www.ag.uiuc.edu/~vista/pdf_pubs/altinsec.pdf

Alternatives to Insecticides for Managing Vegetable Insects by Kimberly A. Stoner. Proceedings from a conference that exchanged experience and research on alternatives to insecticides for vegetable growers in the Northeastern U.S. $8 to NRAES Cooperative Extension, (607) 255-7654; NRAES @cornell.edu; www.nraes.org

A Whole Farm Approach to Managing Pests. Sustainable Agriculture Network (SAN). http://www.sare.org/bulletin/farmpest/. Lays out ecological principles for managing pests in real farm situations. Free in quantity to educators. To order: san_assoc@sare.org; 301-504-5411.

Beneficial Insect Habitat in an Apple Orchard – Effects on Pests. 2004. Research Brief #71. University of Wisconsin-Madison, Center for Integrated Agricultural Systems. http://www.cias.wisc.edu/purr/apples/

Best Management Practices for Crop Pests by Colorado State University Extension. Integrated pest management oriented to western U.S. crops and pests. Bulletin XCM-176. Free ($3 shipping) to Cooperative Extension Resource Center, (877) 692-9358; ResourceCenter@ucm.colostate.edu; www.cerc.colostate.edu

Biodiversity and Pest Management in Agroecosystems (2nd ed.) by Miguel Altieri and Clara Nicholls. Entomological aspects and the ecological basis for the maintenance of biodiversity in agriculture. $49.95 (soft cover) $79.95 (hard cover) from The Haworth Press, Inc., 1-800-HAWORTH; getinfo@ haworthpress.com; www.haworthpress.com

Biological Control in the Western United States. $30 to University of California Press; (800) 994-8849; anrcatalog@ucdavis.edu; www.ipm.ucdavis.edu.

The Control of Internal Parasites in Cattle and Sheep by Jean Duval, Macdonald College, Quebec, Canada (514) 398-7771; eapinfo@macdonald. mcgill.ca; http://www.eap.mcgill.ca/Publications/EAP70.htm

Corn Insect Pests — A Diagnostic Guide. M-166. University of Missouri Cooperative Extension. Service. 1998. 48 p. To order: $8 from (573) 882-7216 or (800) 292-0969. Abstract, summary and/or full version (pdf) online at http://muextension.missouri.edu/xplor/manuals/m00166.htm

Cherry Orchard Floor Management: Opportunities to Improve Profit and Stewardship. MSU Extension Bulletin E-2890. April 2003. http://www.ipm. msu.edu/pdf/E2890CherryReport.pdf

Ecological Engineering For Pest Management: Advances In Habitat Manipulation for Arthropods. G.M. Gurr, S.D. Wratten and M.A. Altieri. 2004. CSIRO Publishing, Collingwood, Australia

Ecological linkages between aboveground and belowground biota. D.A.Wardle, RD Bardgett, JN Klironomos, H Setälä, WH van der Putten et al. 2004. Science 304(11): 1629-1633.

Farmscaping to Enhance Biological Control. 2004. ATTRA. http://attra.ncat. org/attra-pub/summaries/farmscaping.html

Getting the Bugs to Work for You: Biological Control in Organic Agriculture. Symposium Proceedings, November 12, 2004. Portland, Oregon. <http:// csanr.wsu.edu/InfoSources/BugsWorkProceedings.htm>

Guide to the Predators, Parasites and Pathogens Attacking Insect and Mite Pests of Cotton. Publication B-6046. Texas Cooperative Extension. Available on-line or for $5.00 from Distribution and Supply, P.O. Box 1209, Bryan, TX 77806; (888) 900-2577; http://tcebookstore.org/pubinfo.cfm?pubid=748

Habitat management to conserve natural enemies of arthropod pests in agriculture. 2000. Landis, D.A., S.D. Wratten and G.A. Gurr. Annual Review of Entomology 45:175-201.

The Illinois Agriculture Pest Management Handbook. University of Illinois Cooperative Extension Service. 1999. http://www.ag.uiuc.edu/~vista/abstracts/aIAPM.html

Induced Plant Defenses against Pathogens and Herbivores, by Anurag A. Agrawal, Sadik Tuzun and Elizabeth Beth. $59 to American Phytopathological Society Press; (800) 328-7560; http://shop.store.yahoo.com/shopapspress/42422.html.

Insect Pest Management in Field Corn by J. Van Duyn. Discusses cultural practices useful in controlling various insect pests. http://www.ces.ncsu.edu/plymouth/pubs/ent/culprt.html

Integrated Pest Management for Cotton in the Western Region of the United States. Publication 3305. University of California. Division of Agriculture and Natural Resources.1997. Order for $30 from: Agriculture and Natural Resources (ANR) Communication Services, 6701 San Pablo Ave., Oakland, CA 94608; (800) 994-8849 (toll-free order line); (510) 643-5470 FAX; www.anrcatalog.ucdavis.edu

IPM in Practice: Principles and Methods of Integrated Pest Management. $30 from University of California Statewide IPM Program. http://www.ipm.ucdavis.edu/IPMPROJECT/ADS/manual_ipminpractice.html

IPM and Best Management Practice in Arizona Cotton. Part of Publication AZ1006: Cotton, A College of Agriculture Report. College of Agriculture. University of Arizona. 1999. http://ag.arizona.edu/pubs/crops/az1006/az10067a.html

Managing pests with cover crops. Sharad C. Phatak. p. 25–33. In: *Managing Cover Crops Profitably, 2nd Edition.* SAN Handbook Series Book 3. 2001. 212 p. $19 plus $5.95 s/h. http://www.sare.org/publications/

Michigan Field Crop Pest Ecology and Management (E-2704, $12), *Michigan Field Crop Ecology* (E-2646, $12) and *Fruit Ecology and Management* (E-2759, $16) by Dale Mutch et al. Michigan State University Extension. (517) 353-6740; bulletin@msue.msu.edu; http://ceenet.msue.msu.edu/bulletin/sect1021.html

Natural Enemies Handbook: The Illustrated Guide to Biological Pest Control (#3386) by Mary Louise Flint and Steve H. Dreistadt and Pests of the Garden and Small Farm (2nd ed.) (#3332) by Flint. $35 each to University of California Press; (800) 994-8849; anrcatalog@ucdavis.edu; www.ipm.ucdavis.edu.

Natural Enemies of Vegetable Insect Pests by Michael P. Hoffmann and Anne C. Frodsham. $15.70 to Resource Center, 7 Business/Technology Park, Cornell University, Ithaca, NY 14850; (607) 255-2080; http://www.nysaes.cornell.edu/ent/biocontrol/manual.html; resctr@cornell.edu.

Organic Weed Management: A Project of the Northeast Organic Farming Association of Massachusetts by Steve Gilman. NOFA Interstate Council. http://www.nofa.org/

Pest Management at the Crossroads by Charles M. Benbrook. Pest management strategies that rely on interventions keyed to the biology of the pest. $29.95 + $6 s/h. http://www.pmac.net/bymail.htm or (208) 263-5236

The Soil Biology Primer by USDA-NRCS. Describes the importance of soil organisms and the soil food web to soil productivity and water/air quality. http://soils.usda.gov/sqi/soil_quality/soil_biology/soil_biology_primer.html $13 plus $6 s/h to SWCS, (800) THE-SOIL x10; http://www.swcs.org/en/publications/books/soil_biology_primer.cfm

Suppliers of Beneficial Organisms in North America. California Department of Pesticide Regulation. A resource for purchasing biological controls. Free and online in full-text, (916) 324-4247; abraun@cdpr.ca.gov; http://www.cdpr.ca.gov/docs/ipminov/bensuppl.htm

Use of Cultural Practices in Crop Insect Pest Management. E. J. Wright. Nebraska Cooperative Extension. EC95-1560-B. http://ianrwww.unl.edu/pubs/insects/ec1560.htm

Weeds as Teachers: 'Many Little Hammers' Weed Management by Sally Hilander. Proceedings of a 1995 weed management conference on least-toxic and non-toxic techniques for controlling weeds in the Northern Plains (Canada and U.S.). $14 to Alternative Energy Resources Organization (406) 443-7272, aero@aeromt.org; http://sunsite.tus.ac.jp/pub/academic/agriculture/farming-connection/weeds/home.htm

Pests of the Garden and Small Farm: A Growers' Guide to Using Less Pesticide, by Mary Louise Flint. http://www.ucpress.edu/books/pages/2913.html. $19.95.

Biological Control of Insects and Mites: an Introduction to Beneficial Natural Enemies and Their Use in Pest Management, by Daniel L. Mahr and Nino M. Ridgway, University of Wisconsin-Madison. http://muextension.missouri.edu/explore/regpubs/ncr481.htm. To order, request NCR481, *Biological Control of Insects and Mites* ($11). Telephone orders: (573) 882-7216 or (800) 292-0969.

Enhancing Biological Control: Habitat Management to Promote Natural Enemies of Agricultural Pests, edited by C.H. Pickett and R.L. Bugg . http://www.ucpress.edu/books/pages/8180.html. $60.

Conservation Biological Control, edited by Pedro Barbosa. http://www.cplpress.com/contents/C369.htm. £54.95

Handbook of Biological Control, by T. Fisher, Thomas Bellows, L. Caltagirone, D. Dahlsten, Carl Huffake, G. Gordh . http://www.cplpress.com/contents/C360.htm. £110.00

Websites

Agroecology in Action. www.agroeco.org.

ATTRA Pest Management Fact Sheets - http://attra.ncat.org/pest.html

A Whole Farm Approach to Managing Pests. Sustainable Agriculture Network (SAN) - http://www.sare.org/bulletin/farmpest/

Biological Control: A Guide to Natural Enemies in North America, www.nysaes.cornell.edu/ent/biocontrol/

Biological Control as a Component of Sustainable Agriculture, USDA-ARS, Tifton, Ga., http://sacs.cpes.peachnet.edu/lewis

Center for Integrated Pest Management. Technology development, training, and public awareness for IPM nationwide. http://cipm.ncsu.edu/

Database of IPM Resources, a compendium of customized directories of worldwide IPM information resources accessible on line. http://www.ippc.orst.edu/cicp/

Growing Small Farms. Sustainable agriculture website for North Carolina Cooperative Extension, Chatham County Center. http://www.ces.ncsu.edu/chatham/ag/SustAg/index.html

Insect Parasitic Nematodes. The Ohio State University. http://www2.oardc.ohio-state.edu/nematodes/

Iowa State University. http://www.ipm.iastate.edu/ipm/

Michigan State University Entomology http://www.ent.msu.edu/

Michigan State University Insect Ecology and Biological Control www.cips.msu.edu/biocontrol/

North Carolina State University
http://www.ces.ncsu.edu/depts/ent/pestlinks.shtml
http://ipm.ncsu.edu/ncpmip/

Pennsylvania State University IPM, http://paipm.cas.psu.edu

University of California Integrated Pest Management Project, www.ipm.ucdavis.edu

IPM World Textbook, University of Minnesota's list of integrated pest management resources. http://www.ipmworld.umn.edu/

Pest Management at the Crossroads. Comprehensive set of links to ecologically based pest management. www.pmac.net

OrganicAgInfo: http://www.organicaginfo.org/. Has many links to research reports and other publications on pest management and other topics

Regional Experts

These individuals are willing to respond to specific questions in their area of expertise, or to provide referral to others in the pest management field. Please respect their schedules and limited ability to respond. Consider visiting their websites before contacting them directly.

One important source of information is your local Cooperative Extension Service office. Each U.S. state and territory has a state office at its land-grant university and a network of local or regional offices. See http://www.csrees.usda. gov/Extension/ for a listing of all offices.

Northeast Region

Mary Barbercheck
Department of Entomology
516 ASI Building
Penn State University
University Park, PA 16802
(814) 863-2982
(814) 865-3048 – fax
meb34@psu.edu
http.//www.ento.psu.edu/Personnel/Faculty/barbercheck.htm
Soil quality and arthropod diversity as it relates to management of insect pests. Biology and ecology of entomopathogenic (insect-parasitic) nematodes for management of soil-dwelling insect pests.

Brian Caldwell
NOFA-NY Farm Education Coordinator
Northeast Organic Farming Association of New York
education@nofany.org
www.nofany.org
Organic pest management for vegetables and fruit.

Ruth V. Hazzard
University of Massachusetts Amherst
Extension Agriculture and Landscape Program, Vegetable Team
rhazzard@umext.umass.edu
www.umassvegetable.org
IPM, ecological and organic pest management in vegetables.

Fred Magdoff
Department of Plant and Soil Sciences
University of Vermont
frederick.magdoff@uvm.edu
Soil quality, soil testing, ecological soil management.

Ron Prokopy
Department of Entomology
University of Massachusetts
Amherst, MA 01003
prokopy@ent.umass.edu
Ecological management of tree fruit pests

Abby Seaman
New York State Integrated Pest Management Program
Cornell Cooperative Extension
ajs32@cornell.edu
www.nysipm.cornell.edu
Integrating biological controls into vegetable IPM systems

Kimberly Stoner
The Connecticut Agricultural Experiment Station
Kimberly.Stoner@po.state.ct.us
http://www.caes.state.ct.us/
Alternatives to insecticides for managing vegetable insects

John R. Teasdale
USDA-ARS
Beltsville, MD
teasdale@ba.ars.usda.gov
http://www.barc.usda.gov/anri/sasl/sasl.html
Integrated weed management, cover crop management

North Central Region

Dale R. Mutch
Michigan State University
Mutchd@msue.msu.edu
www.kbs.msu.edu/extension
Pest management in farming systems utilizing cover crops

Southern Region

W. Joe Lewis
Research Entomologist
USDA - ARS
Tifton, Georgia
wjl@tifton.usda.gov
http://sacs.cpes.peachnet.edu/lewis/
Sustainable pest management, understanding and enhancing the parasitic and predacious insects that attack plant feeding insects.

David B. Orr
Dept. of Entomology
North Carolina State University
Raleigh, NC 27695-7613
david_orr@ncsu.edu
http://cipm.ncsu.edu/ent/biocontrol/
Biological control of insects in field crops and organic production systems

Sharad C. Phatak, Ph.D.
Professor of Horticulture
100 Horticulture Building
4604 Research Way
University of Georgia
Tifton, GA 31793
(229) 386-3901
(229) 386-3356 – fax
phatak@ uga.edu
Sustainable farming systems, cropping systems, cover crops, conservation tillage and pest management, soil quality and pest management, non-chemical weed management.

Debbie Roos
Agricultural Extension Agent
North Carolina State University
Post Office Box 279 Pittsboro, NC 27312
debbie_roos@ncsu.edu
(919) 542-8202
(919) 542-8246 – fax

http://www.ces.ncsu.edu/chatham/ag/SustAg/index.html
Organic and sustainable agriculture and pest management

Glynn Tillman
USDA-ARS, Tifton, Georgia
pgt@tifton.usda.gov
Biological control of insect pests in cotton

Western Region

Miguel Altieri, Ph.D.
Professor of Agroecology
Division of Insect Biology
University of California, Berkeley
agroeco3@nature.berkeley.edu
www.agroeco.org
Agroecosystem design, biodiversity, ecological pest management

Robert L. Bugg, Ph.D.
Senior Analyst, Agricultural Ecology
U.C. Sustainable Agriculture Research and Education Program
University of California
One Shields Avenue
Davis, CA 95616-8716
(530) 754-8549
(530) 754-8550 – fax
rlbugg@ucdavis.edu
http://www.sarep.ucdavis.edu
*Biological control, on-farm restoration ecology, earthworms, pollinators,
California native plants, cover crops, hedgerows.*

Clara I. Nicholls, Ph.D.
Research Fellow
Divison of Insect Biology
University of California, Berkeley
nicholls@berkeley.edu
www.agroeco.org
Habitat management, biological control

Index

Page numbers in *bold italic* indicate a photo or figure.

organic matter and, 4, 22, 63, 64, 90
pathogens and, 4
pests and, 4
production costs and, 27, 28
soil and, 22, 27, 58–64, 91–92
temperature and, 89
tillage and, 74–75
water and, 22
weeds and, 4, 28, 61–62, 74
yields and, 63, 74
crimson clover, *2*, *33*, *64*
crop health, 103
crop residue, 90
crop rotation
    agrichemicals and, 4, 59, 60–61
    beneficials and, 4, 19
    biodiversity and, 10, 14–15, 16–17, 92, 96–97
    cover crops and, 4, 60–63, 92–93
    no-till and, 60–63, 92–93
    organic matter and, 4, 63
    pathogens and, 4
    pests and, 4
    soil and, 58–59, 62–63, 96–97
    weeds and, 4, 61–62
    yields and, 63
cucumber beetles, *36*, 40

defenses, 44–49
dipteran flies, 79–80
diversity. *See* biodiversity

Ecologically Based Pest Management, 3
ecological pest management
    benefits of, 3
    costs of, 4
    ecosystems and, 1–3, 67
    elements of, 94–95
    goals of, 1–2, 94, 95
    habitat management strategy for, 87–89, 93–95, 100–101
    pillars of, *9*
    whole farm approach of, 18
economic sustainability, 48, 96, 97–99
ecosystems, 1–3, 67, 86
encyrtid wasps (*Encyrtidae*), 83
environment, 1, 48
Epstein, David, 23
erosion, 91, 102
eulophid wasps (*Eulophidae*), 81–82
European mountain ash

(*Sorbus aucuparia*), 34
fairyflies (*Mymaridae*), 81
fertilizers, 64–67, 91. *See also* agrichemicals
filter strips, *34*
flowering plants.
    beneficials and, 12, 19–20, 24, 39, 42, 74–75
    biodiversity and, 12
    corridors and, 37–38
    natural enemies and, 42
    perimeter trap cropping and, 36
    size and shape of, 39
    timing of, 39
Forsline, Philip, *46*
Fort Valley State University, 74
fruit, 47–49. *See also* orchards
fungi, 84. *See also* beneficials; pathogens
fungicides. *See* agrichemicals

genetics, 10–13, 45–46, 84, 100, 101
grapes, 30–33, *46*
green manure, 97. *See also* cover crops
Groff, Steve, 4, 60–63, *62*
ground beetles (*Carabidae*), *38*, 73
gypsy moth caterpillar, *3*

habitat management strategy, 87–89, 93–95, 100–101
hairy vetch, *64*
harvest, 16
health. *See also* resistance
    crop, 44, 103
    soil, 55–58, 89–92, 95, 102
herbicides. *See* agrichemicals
high tunnels, 62
Horton, Dave, 35, *35*

induced defenses, 44
Ing, George, 35
inoculation, 8
insect growth regulators, 59
insecticides. *See* agrichemicals
integrated crop-livestock systems, 5, 15, 16–17, 100
intercropping, 10, 19–20, 89, 92. *See also* corridors; strip-cropping
inundation, 8

Kamut®, 96, 97–98

lacewings (*Chrysoperla* spp.), 73–76

ladybugs (*Coccinellidae*), 7–9, 26, 72–73
lettuce, *37*
livestock, 5, 10, 15, 16–17, 92

manure, 58, 90. *See also* compost; green
    manure
margins. *See* borders
Michigan State University, 20, 22–23
microclimates, 10, 24, 92
minute pirate bugs (*Orius* spp.), 76
mowing, 20, 31–32, 35, 43
mulch, *24*
    agrichemicals and, 60
    beneficials and, 21
    cover crops and, 22, 23
    risks of, 24
    temperature and, 89
mustard, *39*
mycorrhizal fungi, 57

National Academy of Science, 1, 3
natural enemies. *See also* beneficials
    borders and, 25–36
    corridors for, 37–38
    flowering plants and, 42
    overview of, 24, 68
    plant selection and, 39–41
    supplementary resources for, 25
weeds and, 41–43
nematodes, 85. *See also* beneficials;
    pathogens
nitrate leaching, 22
nitrogen. *See* fertilizers
no-till. *See also* tillage
    agrichemicals and, 4, 27, 29, 59–64
    beneficials and, 4, 27–29, 30–33
    cover crops and, 4, 27–28, 30–33, 60–
      63, 92–93
    crop rotation and, 60–63, 92–93
    equipment for, 62–63
    organic matter and, 4, 63, 64
    pathogens and, 4
    pests and, 4
    production costs and, 27, 28
    soil and, 27, 59, 62–63
    weeds and, 4, 29, 61–62
    yields and, 28, 63
Nugent, Mike, 28–29

orchards
    cover crops in, 22–23

margins and, 34
mowing, 35
predators in, 71
resistance and, 47, 48–49
tart cherry orchard, 22–23
organic farming
    Bacillus thuringiensis (Bt) and, 84
    biodiversity and, 87
    economic sustainability and, 96, 97–99
    fertilizers and, 65–67
    pests and, 88
    plant selection and, 88
    weeds and, 99
organic matter. *See also* soil
    aeration and, 57
    beneficials and, 21, 24
    biodiversity and, 86
    compost and, 90
    cover crops and, 4, 22, 63, 64, 90
    crop residue and, 90
    crop rotation and, 4, 63
    fertilizers and, 91
    manure and, 90
    no-till and, 4, 63, 64
    nutrients and, 58
    quality of, 102

parasitoids, 8, 53, 77–83, 88. *See also*
    beneficials
pathogens, 4, 21–24, 83–85. *See also*
    beneficials
perimeter trap cropping, 34–36, 40
pests
    damage caused by, 1
    fertilizers and, 64–66
    habitat management strategy for, 87–89,
      93–95
    key to, 54
    organic farming and, 88
    perimeter trap cropping for, 34–36, 40
    risks of, 25, 26, 90
    soil and, 58–64
plant diversity, 14–15, 18–21, 92–93
plant selection, 39–41, 88, 92, 93
potatoes, *59*
predators, 50–53, 69–77, 88–89. *See also*
    beneficials; natural enemies
production costs, 1, 27, 28, 40, 47
pteromalid wasps (*Pteromalidae*), 82–83
pumpkins, *40*
pyrethroids, 59

Quinn, Bob, 96–99, **98**

rapeseed, **4**
raspberries, 49. *See also* fruit
red clover, **23**
resilience, 13
resistance, 13, 16, 44–49, 58, 64–67, 98
rose clover, **33**
rosemary, **11**
Rosmann, Ron and David, **15**
Rosmann, Ron and Maria, 14–17
rye, **4**, **20**, **59**, **86**

Schomberg, Harry, 74, 75
soil, **56**, **57**. *See also* organic matter
    agrichemicals and, 59–64
    Bacillus thuringiensis (Bt) and, 59
    beneficials and, 21, 58–64
    biodiversity and, 11–13, 15–16
    biological properties of, 55–56, 58
    chemical properties of, 57–58
    compaction, 91, 102
    compost and, 58
    cover crops and, 22, 27, 58–64, 91–92
    crop rotation and, 58–59, 62–63, 96–97
    erosion and, 91, 102
    fertilizers and, 64–67, 91
    health of, 55–58, 89–92, 95, 102
    livestock and, 15
    manure and, 58
    no-till and, 27, 59, 62–63
    pests and, 58–64
    physical properties of, 56–57, 58
    resistance and, 44, 58
    strip-cropping and, 91
    temperature of, 62, 89, 90
    tillage and, 90
    water and, 56–57, 62, 102
Southern green stink bug, **8**
spatial diversity, 10–13, 16–17, 93, 100
spiders, 72
squash, **36**
strip-cropping, 10, 91, 92. *See also* corridors; intercropping
sunflowers, **41**
Sustainable Agriculture Research and Education (SARE), 35, 40, 74
syrphid flies, **25**, 77

tart cherry orchard, 22–23
temperature, 62, 89, 90

temporal diversity, 10–13, 93, 100
Thompson, Dick, 18–19
Thompson, Larry, 12
tillage, 15, 16, 43, 74–75, 90. *See also* no-till
Tillman, Glynn, 74–75
trap crops, 34–36, 40
Trichogramma wasps (*Trichogrammatidae*), 81
Trissolcus basalis, **8**

University of Connecticut, 36, 40
University of Georgia, 59, 74
University of Wisconsin, 48
Upton, Junior, **56**

Vickers, Mark, 27–28
vineyards, 30–33, 37–38, **86**
viruses, 84–85. *See also* beneficials; pathogens

water
    biodiversity and, 10, 92
    cover crops and, 22
    crop residue and, 90
    resistance and, 44
    soil and, 56–57, 62, 102
    wheat and, 98
weeds
    agrichemicals and, 43, 61–62
    beneficials and, 19, 41–43
    biodiversity and, 16
    borders and, 36, 43
    corridors and, 43
    cover crops and, 4, 28, 61–62, 74
    crop rotation and, 4, 61–62
    green manure and, 97
    mowing, 43
    natural enemies and, 41–43
    no-till and, 4, 29, 61–62
    organic farming and, 99
    production costs and, 47
    resistance and, 47
    risks of, 43
    tillage and, 43
wheat, 96, 97–98
Willis, Tad, 96
wind, 14

yields, 28, 40, 63, 74, 98

Zorro fescue, **33**

Books from the Sustainable Agriculture Network

Building a Sustainable Business

A business planning guide for sustainable agriculture entrepreneurs that follows one farm family through the planning, implementation and evaluation process.

280 pp, $17

How to Manage the Blue Orchard Bee

Strategies to rear and manage this alternative orchard pollinator, with details about nesting materials, wintering populations, field management and deterring predators.

88 pp, $9.95

Building Soils for Better Crops

How ecological soil management can raise fertility and yields while reducing environmental impact.

240 pp, $19.95

How to Direct Market Your Beef

Practical tips for successfully selling grass-raised beef to direct markets from one ranching couple's real-life perspective.

96 pp, $14.95

The New Farmers' Market

Covers the latest tips and trends from leading sellers, managers and market planners to best display and sell products. (Discount rates do not apply.)

272 pp, $24.95

Managing Cover Crops Profitably

Comprehensive look at the use of cover crops to improve soil, deter weeds, slow erosion and capture excess nutrients.

212 pp, $19

Steel in the Field

Farmer experience, commercial agricultural engineering expertise, and university research combine to tackle the hard questions of how to reduce weed control costs and herbicide use.

128 pp, $18

The New American Farmer

Profiles 60 farmers and ranchers who raise profits, enhance environmental stewardship and improve the lives of their families.

200 pp, $16.95

Orders

Visit www.sare.org/WebStore to order and preview SAN publications.

Call (301) 374-9696 or send check or money order, specifying title, to:

Sustainable Agriculture Publications
PO Box 753
Waldorf, MD 20604-0753

Add $5.95 for the first book, plus $2 s/h for each additional book shipped within the U.S.A.
MD residents add 5% sales tax. Please allow 3–4 weeks for delivery.

Bulk discounts: Except as indicated above, 25% discount applies to orders of 10–24 titles;
50% discount for orders of 25 or more titles. For more information, see www.sare.org/publications.